CONTEMPORARY CASE STUDIES

2ND EDITION

Population & Migration

Michael Witherick

Series editor: Sue Warn

Philip Allan Updates, an imprint of Hodder Education, an Hachette UK company, Market Place, Deddington, Oxfordshire OX15 0SE

Orders
Bookpoint Ltd, 130 Milton Park, Abingdon, Oxfordshire, OX14 4SB
tel: 01235 827827
fax: 01235 400401
e-mail: education@bookpoint.co.uk

Lines are open 9.00 a.m.–5.00 p.m., Monday to Saturday, with a 24-hour message answering service. You can also order through the Philip Allan Updates website: www.philipallan.co.uk

ISBN 978-1-4441-1982-4

First printed 2011
Impression number 5 4 3
Year 2016 2015 2014 2013

Front cover photograph © Jedi-master/Fotolia

Printed in Dubai

Hachette UK's policy is to use papers that are natural, renewable and recyclable products and made from wood grown in sustainable forests. The logging and manufacturing processes are expected to conform to the environmental regulations of the country of origin.

P01864

Contents

Introduction

Part 1: Population data

Part 2: Population change

Part 3: Population distribution

Part 4: People on the move

Part 5: Impacts of migration

Part 6: Population and resources

Part 7: Population policies

Part 8: Population issues

Part 9: Examination advice

Appendix

Index

Introduction

Much of population geography is about the **distribution** of people on the Earth's surface and the reasons behind this. However, the emphasis is not only on how and why population numbers and densities vary. Population distribution is dynamic; it is changing constantly in response to variations over time in the rate of **natural population change**. Natural population change is the outcome of shifts in the balance between **fertility** and **mortality**, between **birth rates** and **death rates**. There are also changes in population **structure**: for example, in age and gender ratios or in ethnic and household composition. These vary spatially too. Another powerful influence on the changing map of population distribution is **migration**. This is the movement of people over the face of the Earth: for example, from inner city to suburb or from one continent or country to another. International migration is a sensitive issue today, both politically and economically.

The changes described above open up another field of population geography, one that focuses on the impact of people (particularly population growth) on the **consumption** of resources, goods and services. Levels of consumption vary from place to place and have a direct bearing on standards of living and quality of life.

Finally, population geography recognises that governments, together with other decision makers and managers, use their **policies** to try to influence various aspects of population — distribution, change and migration.

About this book

Five main components of population geography are covered in this book. They are:
- change
- distribution
- migration
- resource consumption
- policy

Before looking in turn at each of these components, Part 1 puts the spotlight on a critical aspect of all population geography: the need to have access to reliable and comprehensive data. Because of its contemporary significance, two parts of the book are devoted to migration, the first focuses on motives and the second on impacts.

These five components of population and their key ideas are interlinked — often in a fairly complex way (Figure 1). This should become clear in Part 8 in the analysis of broad population issues which touch all five components. The first eight parts of the book are supported by a range of up-to-date case studies.

Figure 1
The components and key ideas of population geography

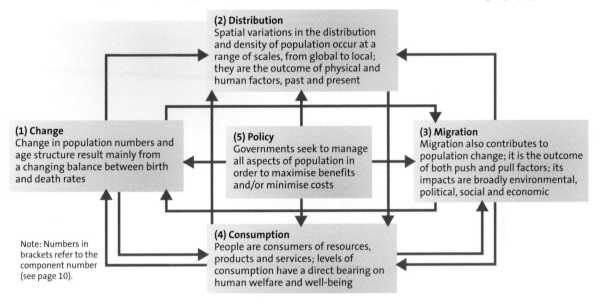

(2) Distribution
Spatial variations in the distribution and density of population occur at a range of scales, from global to local; they are the outcome of physical and human factors, past and present

(1) Change
Change in population numbers and age structure result mainly from a changing balance between birth and death rates

(5) Policy
Governments seek to manage all aspects of population in order to maximise benefits and/or minimise costs

(3) Migration
Migration also contributes to population change; it is the outcome of both push and pull factors; its impacts are broadly environmental, political, social and economic

Note: Numbers in brackets refer to the component number (see page 10).

(4) Consumption
People are consumers of resources, products and services; levels of consumption have a direct bearing on human welfare and well-being

Advice is given throughout the book on making the best use of case studies at both AS and A2. After some of the case studies, there are 'Using case studies' boxes. Most of these show how a particular case study might be useful for answering a specific question. Some invite you to try your hand at an exercise based on one or more of the case studies. You will also find some tips on tackling tasks that are commonly required in examinations, such as:

■ describing the main features of a population distribution map
■ identifying the basic character of a population pyramid
■ extracting the message conveyed by a statistical table
■ analysing a newspaper cutting for bias

The final part of the book consolidates much of this advice and gives some additional tips on making the most of your case study material in the examination.

Key terms

Asylum seeker: a person who seeks to gain entry to another country by claiming to be a victim of persecution, hardship or some other compelling circumstance.

Birth rate: the number of live births per 1,000 people in a year; an indirect measure of a population's fertility.

Carrying capacity: the maximum number of people that can be supported by the resources and technology of a given area.

Consumption: the level of use of resources, goods and services by a population.

Core: an area of concentrated economic development and population.

Counter-urbanisation: the movement of people and employment from major cities to smaller cities and towns as well as rural areas.

Death rate: the number of deaths per 1,000 inhabitants of a given population in a year; a measure of mortality.

Density: the number of people per unit area (usually km^2).

Dependency ratio: the number of children (aged under 15) and old people (aged 65 and over) expressed as a ratio to the number of adults aged between 15 and 64. It indicates the number of people whom the working population has to support.

Distribution: where people are located within a given area.

Ecological footprint: a measurement of the area of land or water required to provide a person (or society) with the energy, food and resources they consume and to deal with the waste they produce.

Ethnic cleansing: a euphemism for the actions of one ethnic or religious group forcing another such group to flee their homes, either by eviction or through fear and intimidation.

Ethnicity: groups of people whose members identify with each other through a common heritage trait such as race, religion and language.

Fertility: the potential of a population to reproduce, often measured as the number of children (under 5 years) per woman of childbearing age (15–50 years) or as the average number of children born per woman.

Genealogy: the study that traces the history or descent of families.

Host area: the destination of a migration.

Human development index (HDI): a measure of national development, which can be used for making international comparisons. The index is based on three equally weighted variables: income per capita, adult literacy and life expectancy. The index takes the lowest and highest values recorded in the world for each variable. The interval between them is given a value of 1 and the figure recorded by each country is then scored on a scale of 0 to 1 (from worst to best). The HDI is the average score of the three variables.

Human trafficking: a modern form of slavery based on an illegal trade in people for the purposes of commercial sexual exploitation or forced labour.

Internally displaced person (IDP): someone who is forced to flee their home but who, unlike a refugee, remains within their country's borders.

Migration: the permanent or semi-permanent movement of people from one place to another. The balance between arrivals in, and departures from, an area impacts on population change.

Mortality: the incidence of death in a population over a given period.

Natural population change: the outcome of the difference between birth and death rates; growth results when births exceed deaths and decline occurs when deaths exceed births.

Optimum population: a theoretically perfect situation in which the population of an area can develop its resources to the best extent and enjoy the highest possible standard of living.

Overpopulation: a situation where the population of an area exceeds its carrying capacity. The symptoms are low (even declining) per capita income and standards of living, unemployment and outward migration.

Periphery: an area lagging in economic development and suffering from net out-migration, high unemployment and a relatively low standard of living; an area losing out to a core.

Policies: the ways in which governments try to manage aspects of the population: for example, change, distribution and migration.

Population change: commonly taken to mean the percentage change in population numbers (i.e. growth or decline) over a year. Strictly speaking, the term includes other shifts over time, particularly in density, structure, migration and consumption.

Refugee: defined by the United Nations as someone whose reasons for moving are genuinely to do with fear of persecution or death.

Remittances: money sent home to family members by migrants working and living abroad.

Replacement rate: the level of fertility at which women are having only enough children to replace themselves and their partner in the population. 'Replacement' is considered only to have occurred when the offspring reach 15 years. In developed countries, a total fertility rate (TFR) of 2.1 is considered to be replacement level (more than 2.0 to allow for childhood mortality).

Source area: where a migration starts from.

Structure: the make-up of a population analysed in terms of, for example, age, gender, marital status, ethnicity, family size and household characteristics.

Total fertility rate (TFR): the average number of children that each woman will have during her lifetime. Theoretically, when the TFR is 2, each pair of parents just replaces itself. In reality, it takes a TFR of 2.1 to 2.2 to replace each generation because some children will die before they grow up to have their own children. A TFR of 2.1 to 2.2 is known as the replacement rate.

Underpopulation: a situation where the resources and development of an area could support a larger population without any lowering of the standard of living, or where a population is too small to develop its resources effectively.

Urban–rural continuum: The almost imperceptible environmental transition from central urban areas to the remote countryside.

Zero population growth (ZPG): the ending of population growth when birth and death rates are equal. This would require a total fertility rate (TFR) of between 2.1 and 2.2.

Population data

Figure 1 (page 7) shows the five components of population geography. Before looking at these in more detail and illustrating them by means of case studies, we need to consider the data that underpin most studies of population and migration. What information is needed and where does it come from?

Referring to Figure 1, reliable statistical information is needed about:

- **population change**: for example, its rate and scale as reflected in birth and death rates, or the numbers of migrants entering and leaving an area (component 1)
- total numbers of people and their distribution at a range of spatial scales from global to local (component 2). Also relevant here would be data about population structure: for example, age, gender, marital status, **ethnicity** and household size
- information about the numbers of migrants leaving and entering an area, their motives for moving and their demographic characteristics, such as age, gender, ethnicity and education (component 3)
- information about the consumption of resources, goods and services, as well as a range of indicators of standard of living and quality of life, such as human welfare, unemployment, per capita income and housing conditions (component 4)

Policy (component 5) is likely to need most of the above statistical information, and more besides.

The census

The main source of such statistical data is the national census. Unfortunately, not all countries have the means to conduct a census at regular intervals because it is expensive and requires a high level of organisation. Even where census data are available, their quality and accuracy vary enormously from country to country. In general terms, it is the developed countries that have the resources and organisational structures to collect large quantities of reliable information. However, the first part of the following case study of the UK's latest 10-yearly census (2001) offers a word of caution. So too does the remainder, but for different reasons.

Case study 1 THE UK'S 2001 AND 2011 CENSUSES

This case study is made up of three parts. The first two look at important outcomes of the 2001 census: namely, a shortfall in the actual number of people recorded and the revelation of a quickening rate of population growth. The third investigates how the 2011 census intends to rectify the first deficiency.

The missing million

The 2001 census put the total population of the UK at 58,789,194. Despite the legal requirement for every householder to complete a census form, inevitably there were a certain number of people who, for various reasons, did not do so. We are told that 94% of the census forms distributed across the country were filled in. However, in the ten boroughs of inner London, for example, the overall completion rate was below 80%; in Kensington and Chelsea it was only 64%. Four per cent of the national forms were subsequently 'estimated' on the basis of a follow-up survey of 300,000 people. The remaining 2%, or about 1 million people, were 'invented' by census officials.

Even so, the total population was still almost a million fewer than had previously been forecast by the registrar general. Census officials believe that the shortfall is partly explained by young men disappearing into the Mediterranean rave culture and by students (mainly male) going around the world on gap-year trips. Apparently, 600,000 more people than had been thought were out of the country. Border control figures collected from around the world suggest that 50,000 were young UK citizens travelling in Australia.

The low response rate, plus the fact that the final population figure was 900,000 fewer than had been predicted, has brought calls for 'a more robust and accurate method of measuring population'. But is there one — will the 2011 census have the answer?

This part of the case study illustrates the point that even in a well-organised developing country, no matter how much care is taken, the results of a census are not always as complete or as accurate as might be thought. Therefore, even official population data need to be handled with some caution.

Population growth: a red light

Despite the missing million, the 2001 census results showed that the UK had the fastest-growing population of any large country in Europe. Its population is now predicted to increase from the present 61 million to 75 million by 2050. As Figure 1.1 shows, that predicted figure is a 'middle of the road' one lying between extreme estimates of 66 and 85 million (for more on population forecasting, see *Case study 5*). Whatever the 2050 total might turn out to be, the important point is that having been stable for several decades, this take-off in the UK's population is ringing alarm bells. In response, the government has commissioned a study to examine the impact on British life, particularly on the quality of life, infrastructure and the environment. This investigation could lead to the adoption of a population policy that tries to slow down the rate of increase.

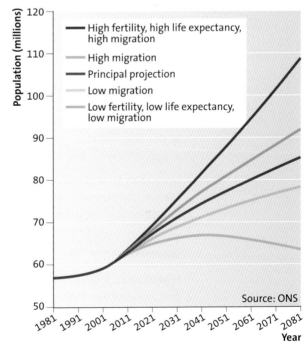

Figure 1.1
Population predictions for the UK, 1981–2081

The last time the government considered the issue of population growth was in 1973, when a study concluded that 'Britain would do better in future with a stationary rather than an increasing population'. The reasons given were that it would help relieve the problems of food supply and an unfavourable balance of payments. As it happened, the government took no action because a falling birth rate, together with low levels of immigration and high levels of emigration, ensured that the UK had a relatively stable population from the mid-1970s to the mid-1990s. Then followed the last Labour government's policy of encouraging immigration to Britain, in order to bridge a skills shortage. This has re-ignited population growth, which is now at its highest rate since the 1970s. The population is increasing by about 250,000 people a year, with about three-quarters accounted for by immigration. Also helping to raise the rate of population growth are the relatively high birth rates among some of the immigrant groups.

This part of the case study makes a number of key points:
- *Close monitoring of population trends is made possible by regular census taking.*
- *Government policy, along with any one of the three components of population change — birth rate, death rate and migration — can have dramatic consequences.*
- *One of the responsibilities of government is to ensure that the national population is kept as close as possible to an optimum level.*
- *How does the UK's recent experience fit the demographic transition model? Is the UK regressing to Stage 3 or is it defining a new Stage 5 (see* Case study 7)?

Tracking down the 'missing' people

In advance of the next UK census to be held in April 2011 the Office for National Statistics (ONS) has been working with local authorities and community groups to make sure that it accurately records the diversity of population in the UK. Special attention has been paid to those hard-to-count groups which, it is believed, were under-counted in the previous census. These groups include:
- disabled and very elderly people
- ethnic minority groups
- faith communities
- migrants
- non-English-speaking people
- unemployed people
- people on low income
- students and other young adults
- gypsy, traveller and other groups where response has been historically low

Making an accurate count of these groups is clearly going to be a major challenge. It will involve the expenditure of a considerable amount of time and money. What is to be gained from these data?

The question for you to consider is whether or not you feel that the expenditure is justified. What is to be lost if these groups are under-enumerated?

Newsflash: It looks as if the 2011 census may be the UK's last. It is estimated that it will cost around £480 million. The new coalition government of 2010 claimed that this was too high a price to pay for data that are so quickly out of date in our fast-moving world. It suggested that there may be other sources from which to draw required demographic data. In future, data could be gathered instead from records held by the Post Office, HM Revenue and Customs (formerly Inland Revenue), local government and credit-checking agencies.

ICT and population data

We live in an era of information and communications technology (ICT). What this means in the context of population is that we are now able to collect and store much more data about people. The population database is no longer limited to numbers and where people happen to live. It can embrace a whole range of detailed information about people's demographic characteristics. These include their gender, age, ethnicity, class and so on. But ICT does not end with the collection and storage of data. It also includes the retrieval of those data and their conversion into a digital format, together with their manipulation and analysis by computers and a growing range of digital technologies. Then there is the vitally important communications component of ICT. In other words, there is the transmission and receipt of digital data in either a 'raw' or a 'processed' format.

WHAT'S ON THE INTERNET?

Case study 2

The internet is the highway along which people and businesses transmit and receive mind-blowing amounts of Information. Table 1.1 illustrates the speed at which the world as a whole became users of this highway during the first decade of the twenty-first century. From a small base in 2000, Africa, the Middle East and Latin America have shown remarkable percentage rises in the number of internet users. North America is well clear of the other global regions in terms of the percentage of the population using the internet.

Table 1.1
World internet usage, 2000–09

World regions	Population, 2009 (est.)	Internet users, 2000	Internet users, 2009	Penetration (% of population)	Growth in internet users, 2000–09	Users (% of world total)
Africa	991,002,342	4,514,400	86,217,900	8.7%	1,809.8%	4.8%
Asia	3,808,070,503	114,304,000	764,435,900	20.1%	568.8%	42.4%
Europe	803,850,858	105,096,093	425,773,571	53.0%	305.1%	23.6%
Middle East	202,687,005	3,284,800	58,309,546	28.8%	1,675.1%	3.2%
North America	340,831,831	108,096,800	259,561,000	76.2%	140.1%	14.4%
Latin America/ Caribbean	586,662,468	18,068,919	186,922,050	31.9%	934.5%	10.4%
Oceania/Australia	34,700,201	7,620,480	21,110,490	60.8%	177.0%	1.2%
World total	6,767,805,208	360,985,492	1,802,330,457	26.6%	399.3%	100.0%

Source: Internet World Stats

The internet allows users to access a vast amount of population data. Reliable international compilations of data can be obtained from a number of organisations, such as the World Bank (http://data.worldbank.org/topic), the United Nations (http://unstats.un.org), the US Central Intelligence Agency (CIA) (www.cia.gov/library/publications/the-world-factbook) and the US Census Bureau (www.census.gov/ipc/). These same websites can also supply statistical summaries for all the countries of the world. More detailed national data are to be found on the respective government websites. For example, for the UK it is the Office for National Statistics (www.statistics.gov.uk); for the USA, the US Census Bureau (as above); for China, the National Bureau of Statistics of China (www.stats.gov.cn/english); and for Ethiopia, the Ethiopian Central Statistical Agency (www.csa.gov.et). Basically, the quality of data available on national websites such as these depends very much on the rigour of the national census.

Also available on the internet, but in most cases for a fee, are the services of data analyst companies and organisations. Examples are Market Research.com, which offers a global service, and nation-specific organisations such as Indiastat.com and Instituto Brasileiro de Geografia e Estastica (IBGE).

Up-to-date maps showing population densities and distribution at a global level are readily available. Less commonly encountered are maps showing spatial variations in population within individual countries. The UK's Office for National Statistics (see above) allows access to neighbourhood statistics. Using these data, it is possible to map the distributions of various demographic attributes for, say, a town or city. The US Census Bureau has a website (www.census.gov/ipc/www/idb/informationGateway.php) which will construct the population pyramid for any country and for any year from 1950 onwards (see *Case study 3*).

There is an almost limitless supply of population data to be found on the internet, particularly at international and national levels. Users of such data need always to be aware of their sources and to check their reliability. Inaccuracy and bias are surprisingly common flaws.

1

Write a short account analysing the data given in Table 1.1.

Guidance

A purposeful analysis might focus first on the percentage growth in internet users (column 6), then the degree of penetration (column 5) and finally put the regions in a global context by looking at their percentage shares of all internet users (column 7). To put the last into a proper perspective, it would be important to be aware of differences in the regional populations (column 2). You should also be alive to the fact that Asia is such an immense region that it embraces striking contrasts in internet usage at an international level.

Using case studies

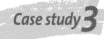

Case study 3 **READY-MADE POPULATION PYRAMIDS**

The population pyramid is one of the most important tools used in the analysis of population. Basically the pyramid unravels the structure of a population in terms of gender and age. It is in fact a histogram, constructed in 1-, 5- or 10-year age groups, with males on one side, females on the other. The base of the pyramid represents the youngest age group and the apex the oldest. The horizontal bars are drawn proportional in length to either the percentage of the total population (a format favoured by the US Census Bureau — see below) or the actual number in each age group.

Two websites in particular are to be recommended. The US Census Bureau's International Data Base website (www.census.gov/ipc/www/idb) allows you to view and print off the population pyramid of any country. Not only that, but you can access the pyramid for any year from 1950 to the present, and indeed predicted pyramids up to 2050. So the pyramids available on this website facilitate three types of demographic investigation:

- population structure at a particular point in time
- change in population structure over a period of time
- international comparisons, as for example in the context of the demographic transition (see Part 2, pages 24–26)

But the interest in population structure is not confined to the national level and international comparisons — important though they might be. The variations in population structure within a country are of equal interest. This is where the UK's Office for National Statistics (www.statistics.gov.uk/census2001/pyramids/pages) comes into its own. It is possible to download population pyramids of all the countries and regions of the UK and their constituent local government areas.

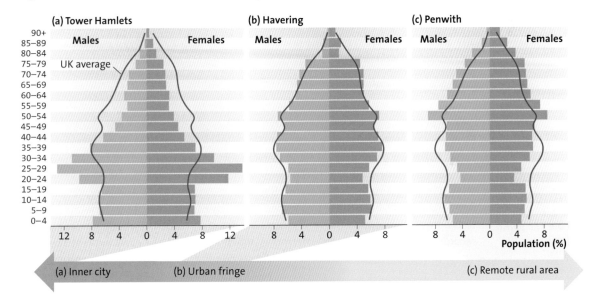

The three population pyramids in Figure 1.2 illustrate how population structure changes as one moves along the **urban–rural continuum** from the city centre to the remote countryside. Any comparative analysis of population pyramids is enhanced if there is a common datum, such as the precise shape of the national population pyramid for the same year. Analysis may then be focused on how each pyramid deviates from the national norm.

So what might we extract from the three pyramids in Figure 1.2?

- Pyramid A is for the London borough of Tower Hamlets (inner city) — above-average numbers of young adults taking advantage of the flats and terraced housing close to the employment opportunities of central London. Below-average incidence of older age groups (> 40 years) reflects (i) the outward movement of households with children to larger dwellings in the suburbs, (ii) retirement moves to more pleasant

Figure 1.2
Changing population structure along the urban–rural continuum in the UK

residential locations and (iii) possibly lower life expectancies. The base of the pyramid, although undercut, suggests birth rates slightly above the national average. This may reflect the presence of (i) relatively large numbers of young adults and (ii) immigrants from the Indian subcontinent.

- Pyramid B is for the London borough of Havering (urban fringe) — the remarkable feature of this pyramid is the closeness with which its overall shape matches the national average. It shows a slightly above-average presence of people aged > 50 years. This might be the outcome of retirement moves and slightly raised life expectancies related to the more attractive and healthy environment to be found on the urban fringe.
- Pyramid C is for Penwith in West Cornwall (remote rural) — the pyramid is top heavy with an under-representation of young people (aged < 44 years) and an over-representation of old people (aged > 50 years). The former reflects the lack of employment opportunities and an outward migration of young adults in search of work. The latter is the outcome of the popularity of this area as a place of retirement — a combination of a pleasant environment, mild winters and relatively cheap property.

The population pyramid is a much-used diagram in population studies. Not only does it give a clear visual analysis of the age and gender structure of today's population, but it also provides a valuable tool in predicting the size and structure of future populations.

For more on population pyramids, see Using case studies 2 *on page 25.*

Case study 4 — RESEARCHING YOUR FAMILY TREE

Today there is a growing interest in family roots and reconstructing family trees that represent a family's history. This interest in **genealogy** has no doubt been kindled by television programmes, such as the BBC's *Who Do You Think You Are?* Researching family history has been made much easier made by the internet. It has facilitated access to the three different demographic facts needed to reconstruct a family tree: namely, records of births (or baptisms), marriages and deaths. From the early sixteenth century to the mid-nineteenth century, this information was painstakingly collected and recorded in almost every British parish (Figure 1.3). Sometimes, there is also reference to the occupations of individuals. Sadly, over the centuries some of these parish records have been lost or destroyed. Nonetheless, enough of them survive to provide an invaluable source of data.

Figure 1.3
Page out of a parish register

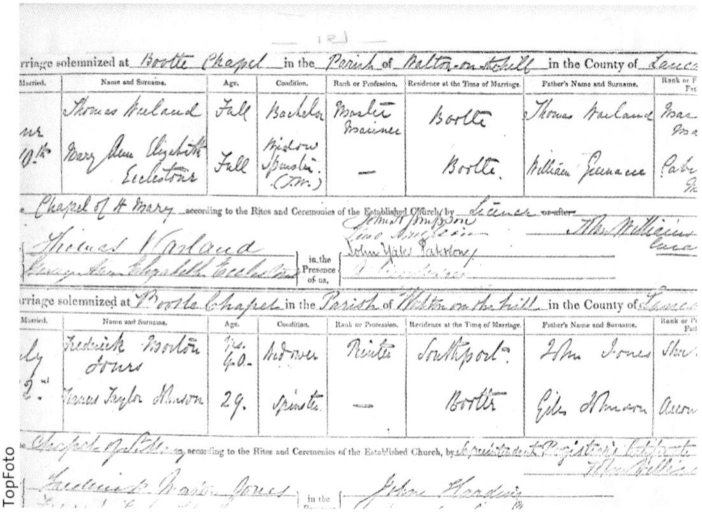

TopFoto

The civil, as opposed to the ecclesiastical or parish, registration of births, marriages and deaths started in 1837 in England and Wales. This was expanded in 1927 to include stillbirths and adoptions. The registration system persists to

Contemporary Case Studies

this day. All the records are kept at the General Register Office. For a fee, it is possible to gain information about the birth, marriage and death of an individual and where these registrations were made.

Another important data source is the UK's 10-yearly censuses. The enumerator's returns record individual people by name, gender, age and occupation on a house-by-house, street-by-street, and settlement-by-settlement basis (Figure 1.4). Census records for England and Wales from 1841 to 1911 are available online (www.nationalarchives. gov.uk/records/census-records.htm). Searching the websites involved is free, but there is a charge to download documents. Most local and county record offices hold microfilm or microfiche copies of the 1841 to 1911 **census returns** for their own areas. The enumeration records of 1921 and all later censuses remain in the custody of the Office for National Statistics until 100 years have elapsed since the actual date of the particular census. It is only then that the records become part of the National Archives and accessible to the general public.

Other sources of information include gravestone inscriptions, obituaries, tax lists, wills and other miscellaneous types of record. With the exception of the first, these data sources are most likely to be found in county record offices, libraries or museums.

Figure 1.4
Completing a census return

So great is the interest in family histories today that there are now available many software packages (e.g. Family Tree Maker, Legacy and Ancestral Quest) and dedicated websites (e.g. www.findmypast.co.uk, www.family-historian.co.uk and www.rootsmagic.com). Both types of source make the accessing of data and the whole research procedure much easier, but they both cost money. Reconstructing a family tree can become an expensive interest.

So why are people prepared to spend money as well as time researching their family tree? For many, it is nothing more than an absorbing leisure activity. Most of us have an innate curiosity about our roots. But there is also an academic interest because collective family histories can inform us on a range of demographic topics:

- family size
- life expectancy
- migration moves
- social mobility

Individual family histories can add a touch of realism to rather detached generalisations about population change. This is particularly so if the detail of the family tree is set in the context of a country's social and economic history.

Population change

Global population growth

During 2009, at least another 65 million people were added to the global population. At the end of the year, that total population was estimated to be 6.8 billion. In percentage terms, the rate of population growth in 2009 was less than during the 1980s and 1990s. Global population growth rates have fallen from 2.1% per year to around 1.95%. Besides showing the upward curve of global population since 1800, Figure 2.1 shows how long it has taken for the world population to increase by 1 billion. It took 118 years from 1804 for the population to rise from 1 to 2 billion. Since then the length of time to add another billion has shortened to a mere 12 years, between 1987 and 1999. What Figure 2.1 also shows is that the length of time is just beginning to increase, from 12 years to 14 years. In short, it looks as if the rate of global population growth may be beginning to slow down.

The interesting question is: given this slowing down, when exactly will the point of **zero population growth (ZPG)** be reached? There is much disagreement about the likely date. Some have suggested that it might be as early as 2040. Others say that this is far too optimistic and that the threshold will not be reached before 2080 (Figure 2.2). Reality may well lie somewhere between those two dates.

Forecasting future population growth is a tricky business. The problem is that so often there are surprises or unforeseen developments. For example, population

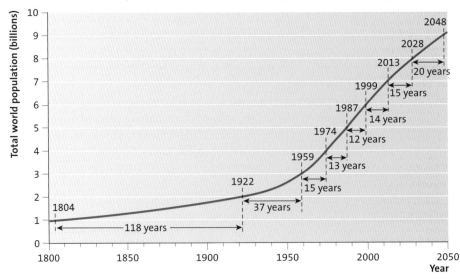

Figure 2.1
Global population growth rates, 1800–2050

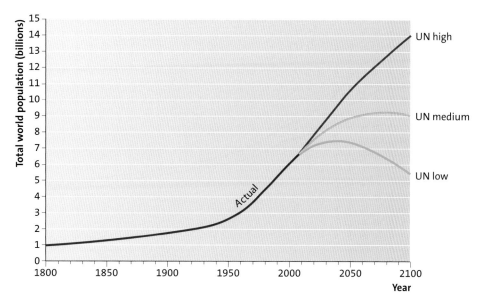

Figure 2.2
Global population growth, 1800–2100

growth in developing countries was lower than expected during the 1990s. In the UK, birth rates have increased since 2000 and this has caused the population to surge past the 60 million mark (see *Case study 1*). However, such surprises do not seem to deter demographers from making predictions about, say, when the global population will peak and at what total figure. But rather than making just one prediction, it is common practice for a number of predictions to be made for a range of different future scenarios. Each scenario will make specific assumptions about rates of fertility and mortality (Figure 2.2). This multi-prediction approach is also illustrated in Figure 1.1 (on page 11) with respect to the UK's future population. The beauty of such an approach is that subsequent reality will often be found to lie somewhere between the extremes of 'high' and 'low' forecasts.

PREDICTING GLOBAL POPULATION GROWTH

Case study 5

Although most demographers are agreed that there will come a time in the Earth's history when the global population ceases to grow, no one knows for certain when this will be. What actually happens to population growth depends on a number of factors. Some of these can be estimated with confidence, while some cannot. Two that can are:
■ the age structure of the population
■ the **total fertility rate (TFR)**

The TFR is the average number of children that each woman will have during her lifetime. It is an average because, of course, some women will have more, some fewer and some no children at all. Theoretically, when the TFR is 2, each pair of parents just replaces itself. In reality, it takes a TFR of 2.1 to 2.2 to replace each generation because some children will die before they grow up to have their own children. A TFR of 2.1 to 2.2 is known as the **replacement rate (RR)**.

   In 2010 the global TFR stood at 2.56 in 2010. It has been declining since 1950 when it was a staggering 5.01. Much of this decline in TFRs has taken place in developing countries, presumably as the various factors involved in the demographic transition take

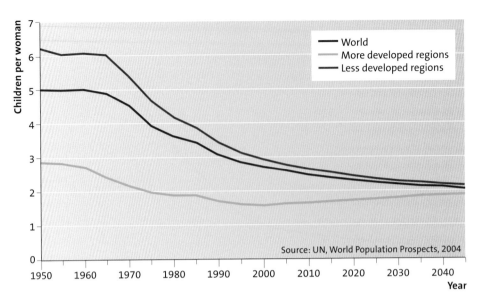

Figure 2.3
Global total fertility rates, 1950–2045

Source: UN, World Population Prospects, 2004

hold. The UN now estimates that the TFR will reach the RR threshold in 2040 (Figure 2.3). Even so, it is predicted that it will be some time after that before **zero population growth (ZPG)** is reached. However, even the smallest shift in the TRF can have an immense impact on future population. For example, while a TFR of 2.06 is predicted to lead to a stable population of 11 billion by 2075, a rate of 2.17 would produce a population of 20 billion that will continue to rise well beyond that year.

The several agencies whose task it is to predict future population seem to be moving closer to a consensus that:

- global population will continue to grow after 2050
- it will peak a little over the 9 billion mark (more than 2 billion people than today) some time during the second half of the twenty-first century (Table 2.1)

Table 2.1
UN (2008) and US Census Bureau (2010) estimates of global population growth, 2010–50

Year	UN 'medium' estimate (millions)	Increase	US estimate (millions)	Increase
2010	6,909		6,831	
2020	7,675	766	7,558	727
2030	8,309	634	8,202	644
2040	8,801	492	8,749	547
2050	9,150	349	9,202	453

Despite this consensus, there remains a degree of uncertainty and a great deal of concern. What is the world's ultimate **carrying capacity**? Will it be able to cope with the needs of 9 billion people (see *Case study 25*)? What will happen to the world if the eventual peak population greatly exceeds 9 billion?

The key point of this case study is this. It is vitally important that we try to forecast future population levels, if only to ring alarm bells or to ensure that we are properly prepared to cope with predicted demographic scenarios. However, population forecasting is a fraught business complicated by the range of factors to be taken into account and second guessing many unknowns. Inevitably, it is necessary to work within the tramlines of pessimistic and optimist forecasts.

Contemporary Case Studies

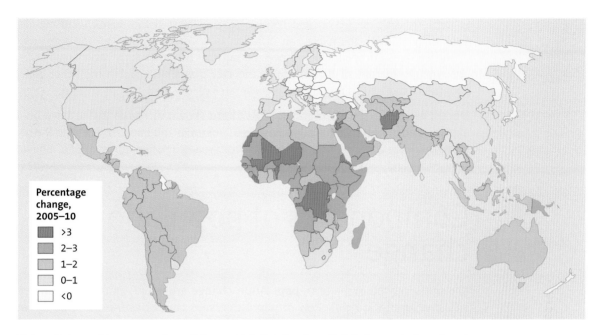

Percentage change, 2005–10

- >3
- 2–3
- 1–2
- 0–1
- <0

Figure 2.4 provides a snapshot of the spatial distribution of global population change during the first decade of the twenty-first century. Catching the eye are the high rates of population growth (over 2% per annum) over much of Africa and parts of the Middle East. Lesser growth rates of between 1 and 2% are seen as occurring over much of the globe — Latin America, the extremes of north and south Africa, and southern and southeast Asia and Australasia. In other words, included here are large tracts of the developing world, as well as a few developed countries.

The developed world is seen to be divided between those parts showing little positive change (less than 1% per annum), such as western Europe and North America, and those parts declining in population, such as eastern Europe, the Russian Federation and Japan. It is interesting to note that 12 of the countries listed in Table 2.2 belonged

Figure 2.4
The global distribution of population change, 2005–10

Table 2.2
Countries with negative or zero population growth (2010)

Country	Annual rate of natural change	Predicted % decline in population by 2050	Country	Annual rate of natural change	Predicted % decline in population by 2050
Ukraine	−0.8	28	Croatia	−0.2	14
Russia	−0.6	12	Germany	−0.2	9
Belarus	−0.6	12	Czech Republic	−0.1	8
Bulgaria	−0.5	34	Japan	0	21
Latvia	−0.5	23	Poland	0	17
Lithuania	−0.4	15	Slovakia	0	12
Hungary	−0.3	11	Austria	0	8
Romania	−0.2	29	Italy	0	5
Estonia	−0.2	23	Slovenia	0	5
Moldova	−0.2	21	Greece	0	4

Source: Population Reference Bureau

to the former Soviet Union and its bloc of satellite states in eastern Europe (see *Case study 8*). With the exception of Japan (see *Case study 28*), all the remaining countries are European developed countries. A particularly noteworthy and encouraging feature of Figure 2.4 is the fact that China and some other Asian countries experienced population growth rates of less than 1% per annum.

Modest rates of population growth (1–2%) are shown in Figure 2.4 to have been experienced in India, China, southeast Asia, Australia and Latin America. In the last of these areas, there were a few 'islands' of greater change (for example, Bolivia, Guatemala and Nicaragua).

Components of population change

The shifts implied in the term **population change** are mainly to do with numbers and structure. A systems view of population sees change as the outcome of two processes — **natural change** and **net migration**. The inputs are births and inward migration (**immigration**), while the outputs are deaths and outward migration (**emigration**) (Figure 2.5).

- Natural change reflects the balance between births (birth rates, fertility) and deaths (death rates, mortality). Natural increase results when births exceed deaths; natural decrease occurs when deaths exceed births.
- Net migration or migrational change can also be either positive or negative depending on the balance between arrivals and departures.

The nature of the balance between natural change and net migration determines both the direction and rate of population change. Broadly speaking, these two

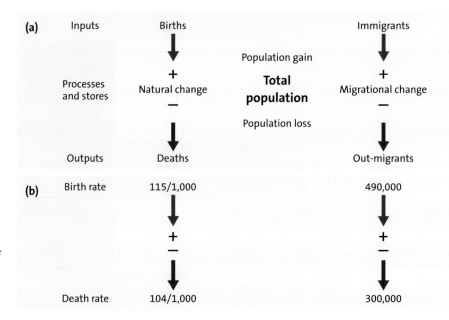

Figure 2.5
(a) A systems view of population.
(b) The UK population system, 2001

factors can work either together or against each other (Figure 2.6). If they work together, the level of population can rise or fall significantly. If they are in opposition, one of the components will tend to neutralise the impact of the other. Therefore, the result will be a reduction in the potential scale of population change, be it plus or minus.

The global map of population change (Figure 2.4) is affected most of all by the balance between birth and death rates. To explain the global distributions of each of these, a whole range of factors needs to be considered (Figure 2.7).

Changes in population **structure** are also important. Of the various attributes of structure, age is probably the most significant. The age structure of a population has a direct bearing on birth and death rates and therefore on natural change (see *Using case studies 2*). The population or age–sex pyramid perhaps best shows the link between age structure and population change (Figure 2.8). (The demographic transition model is explained later in this section.)

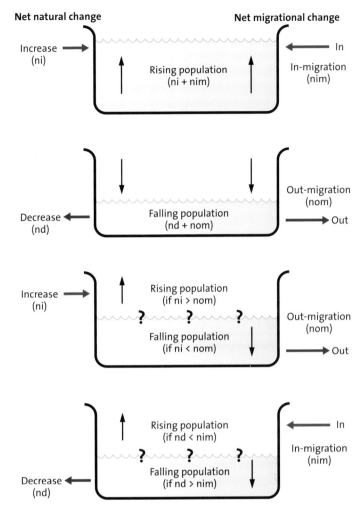

Figure 2.6
Different population change scenarios

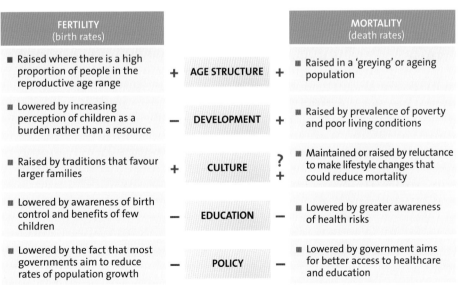

FERTILITY (birth rates)			MORTALITY (death rates)
■ Raised where there is a high proportion of people in the reproductive age range	+ AGE STRUCTURE +		■ Raised in a 'greying' or ageing population
■ Lowered by increasing perception of children as a burden rather than a resource	− DEVELOPMENT +		■ Raised by prevalence of poverty and poor living conditions
■ Raised by traditions that favour larger families	+ CULTURE	? +	■ Maintained or raised by reluctance to make lifestyle changes that could reduce mortality
■ Lowered by awareness of birth control and benefits of few children	− EDUCATION −		■ Lowered by greater awareness of health risks
■ Lowered by the fact that most governments aim to reduce rates of population growth	− POLICY −		■ Lowered by government aims for better access to healthcare and education

Figure 2.7
Some factors affecting fertility and mortality

Figure 2.8
The demographic transition model applied to the UK

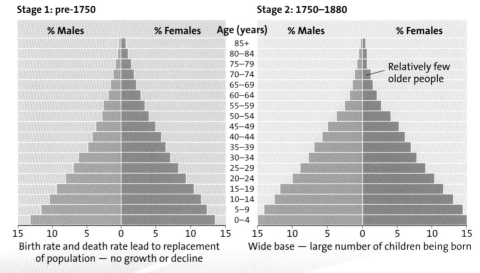

Stage 1: pre-1750

% Males   % Females   Age (years)

Birth rate and death rate lead to replacement of population — no growth or decline

Stage 2: 1750–1880

% Males   % Females

Relatively few older people

Wide base — large number of children being born

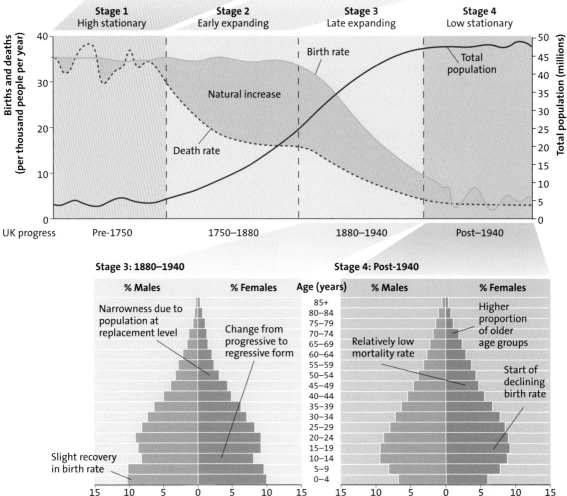

| | Stage 1
High stationary | Stage 2
Early expanding | Stage 3
Late expanding | Stage 4
Low stationary |

Birth rate

Total population

Natural increase

Death rate

Births and deaths (per thousand people per year)

Total population (millions)

UK progress   Pre-1750   1750–1880   1880–1940   Post–1940

Stage 3: 1880–1940

% Males   % Females   Age (years)

Narrowness due to population at replacement level

Change from progressive to regressive form

Slight recovery in birth rate

Recent growth

Stage 4: Post-1940

% Males   % Females

Higher proportion of older age groups

Relatively low mortality rate

Start of declining birth rate

Narrow base — fewer children being born

A broad-based pyramid tapering fairly abruptly upwards (Stage 2, Figure 2.8) indicates a youthful population. High rates of fertility and natural increase are to be expected, given the large proportion of people of reproductive age. Conversely, a pyramid shaped like a beehive (Stage 4) indicates an ageing or 'greying' population. Low fertility is reflected in the undercutting of the base, while the gentle taper is the product of long life expectancy and low mortality. In Stage 3, the much narrower base signals a decline in fertility, while the gentle upward taper again suggests that mortality is quite low. A slower rate of population increase is to be expected. Fertility and mortality are so balanced in Stage 1 that a significant change in population numbers is highly unlikely.

2 *Using case studies*

Coping with population pyramids

Population pyramids occur frequently in examination questions on population geography. Candidates are asked to complete any one of at least four different tasks:

- annotate a pyramid to draw attention to the main features of the population
- suggest what sort of country is being represented by a particular pyramid — roughly where is it located along the demographic transition model (DTM)?
- compare the pyramids of two or more populations
- relate pyramids to different stages in the demographic transition model

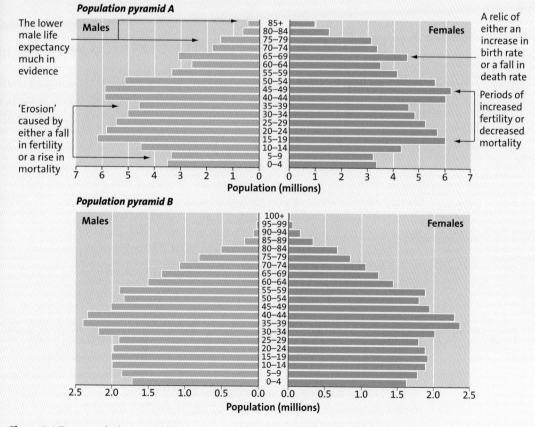

Figure 2.9 *Two population pyramids*

The demographic transition

The **demographic transition model (DTM)** puts the changes in population numbers, birth and death rates, and population structure into a single time sequence (see Figure 2.8). In the model, countries make the transition from a situation where high mortality in a youthful population produces little growth to one where a similar situation results from low fertility in an ageing population. In between these two states, birth and death rates at first diverge and then converge, producing phases of accelerating and then decelerating growth. These are the critical phases. The model usually recognises four stages, but events in the UK during the last intercensal period (1991–2001) (see *Case studies 1* and *7*) lead us to wonder whether the model might need updating and extending to include a fifth stage (contracting, or under-populating). On the other hand, recent developments in Russia (see *Case study 8*) and other countries shown in Table 2.2 perhaps suggest the onset of a rather different scenario (see *Case study 28*).

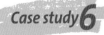 *Case study 6* THREE ASIAN COUNTRIES: DO THEY FOLLOW THE DTM?

It is often claimed that there is a close relationship between the development process and the progress made by countries through the demographic transition. The UK was perhaps the first country to go through the transition (Figure 2.8). The driving force was the industrial revolution, which began during the second half of the eighteenth century. By about 1940, the UK had reached the start of Stage 4, having completed Stages 2 and 3 in a little under 200 years (see Figure 2.10 (a)).

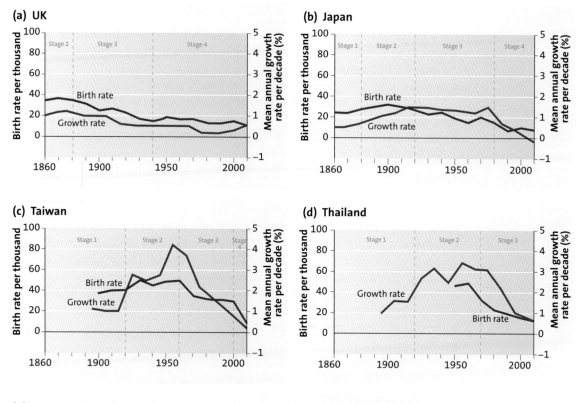

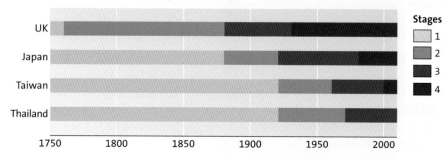

Figure 2.10
Four demographic transitions

The aim of this case study is to examine how three countries in Asia have fared in terms of the demographic transition. Four questions will be addressed:

- Do these countries appear to have gone through, or to be going through, the same sequence of stages?
- If 'yes', how far have they progressed along the transition?
- How does their speed of progress compare with that of the UK?
- What links, if any, can be found between the DTM and economic development?

The chosen Asian examples are Japan, Taiwan and Thailand. Birth rates and mean annual growth rates are used to compare and monitor their progress along the DTM (Figure 2.10).

Japan

The industrialisation and urbanisation of Japan began around 1860 (some 100 years after the UK). A little over 100 years later, Japan had become an economic superpower.

Figure 2.10 (b) suggests that Japan:

■ deviated slightly from the DTM with its steady rate of population growth during Stages 2 and 3 (i.e. between 1880 and 1980); growth appears to have been fuelled by a high birth rate

■ progressed through Stages 2 and 3 in about half the time it took the UK and has now reached the point of ZPG (Table 2.2)

Taiwan

The development process in Taiwan began fairly slowly at the beginning of the twentieth century. At this time, it was a Japanese colony. After the Second World War, thanks to support from the USA, development was based on agriculture and a range of export-oriented light industries. Between 1960 and 1990, the economy grew at an annual average rate of 9.5% and Taiwan joined the ranks of the so-called 'Asian Tigers' — a group of newly industrialising countries (NICs). At present, the country is somewhat in the doldrums. Like others it is suffering from the slump in the global economy. Figure 2.10 (c) suggests that:

■ the critical Stage 2 started around 1920
■ a high birth rate persisted to around 1960
■ the onset of Stage 3 was marked by the spectacular plunge in the birth rate and in the mean annual rate of population growth between 1960 and the 1990s
■ with the mean annual growth rate down 0.2% per annum Taiwan looks to have entered Stage 4

Thus it seems that not only has Taiwan been running some 40 years behind Japan, it has taken slightly less time to complete the two critical stages of the transition.

Thailand

Thailand belongs to the next generation of developing nations, presently known as the recently industrialising countries (RICs). Between 1920 and 1970, the main feature of economic development was a steady rise in agricultural productivity. During the 1970s, the economy broadened to industries and services. The period from 1985 to 1995 was very prosperous; the average annual growth rate was more than 7.5%. In 1997, the economic bubble burst and this, together with the spread of 'Asian flu', put Thailand into a recession from which it is only just beginning to recover.

Thailand's economic development has not only come later than that of Japan and Taiwan; it also differs in terms of the relative importance of agricultural exports and tourism.

Figure 2.10 (d) suggests that:

■ Thailand possibly started Stage 2 around 1920 (the same time as Taiwan), but the lack of birth-rate data makes this uncertain
■ it was not until around 1960 that the growth rate began to fall from a high level of 3% per annum, thus marking the graduation from Stage 2 to Stage 3
■ although there are clear signs that the rate of population growth is continuing to fall, it cannot be said for sure that the country has yet reached Stage 4

Thailand appears to have moved slightly more slowly along the demographic transition than Taiwan. In both instances, the graduation from Stage 2 to Stage 3 seems to have coincided with a significant take-off in economic growth — during the 1960s in Taiwan and during the 1970s in Thailand. Equally, the move towards Stage 4, as evidenced by the flattening out of population growth rates, looks to have possibly been hastened by the global economic recession of the late 1990s that hit Asian countries so badly.

3

Using case studies

Here is a task for you to try yourself.

Question

Read Case study 6, paying particular attention to Figure 2.10 (especially (e)).
(a) Do all four countries appear to have gone through, or to be going through, the same broad sequence of demographic change?
(b) How do their rates of progress through the model compare?
(c) How might you explain any differences in the rate of progress?
(d) What demographic change appears to mark the beginning of:
  (i) Stage 2?
  (ii) Stage 3?
(e) Explain the possible links between the demographic transition and economic development.

A NEW STAGE IN THE DTM (UK)

Case study 7

Between 1870 and the beginning of the Second World War in 1939, the UK's birth rate fell from 37 to 12 births per 1,000 population, so taking the country into Stage 4 of the demographic transition (see Figure 2.8). That fall can be explained in terms of:
- the increased use of birth control methods
- more employment opportunities for women
- the rise of an urban society that no longer saw any advantage in having large numbers of children
- the growing perception of children as a drain on household finances and an impediment to socioeconomic progress

Figure 2.10(a) shows that throughout the second half of the twentieth century, the birth rate fluctuated between 13 and 18 per 1,000. At the last census (2001), it stood at 12 per 1,000. Is the persistence of low fertility in the UK still explained by the same four factors?

The short answer is that those explanatory factors remain valid and may even have been strengthened. For example, more reliable and readily available contraception means more effective birth control. British society has become even more urbanised, and the costs of housing and child-raising have rocketed. For increasing numbers of people, a career comes before children. However, as Figure 2.11 shows, these are not the only factors involved.

The progressive 'ageing' of the UK's population is reducing the percentage of the population that is of reproductive age, so falling birth rates set in motion a downward spiral of even lower birth rates (Figure 2.11). The second additional factor is a set of

Figure 2.11
Factors reinforcing the persistence of low birth rates

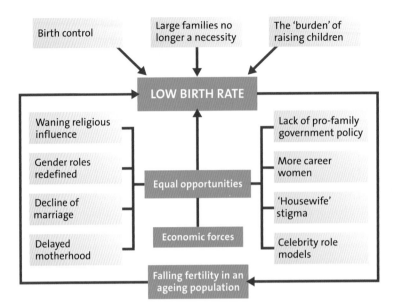

changes in British society linked, to varying degrees, with legislation put in place over the last 50 years to end sex discrimination and enforce equal opportunities. These changes are outlined below:

- **gender roles redefined** — for example the advent of working women and 'house husbands'
- **declining incidence of marriage** — the increasing popularity of couples living together outside marriage and without children
- **delayed motherhood** — many women who choose to have children are postponing their first pregnancy until they are in their thirties and have had time to forge a career; this often results in one-child families
- **economic pressures** — the high costs of housing, living and raising children
- **government policy** — little pro-family emphasis
- **more career women** — increasing numbers of women are choosing not to marry or to have children
- **the 'full-time housewife' stigma** — women complain that their peers look down on them if they give up their careers and stay at home to raise children
- **celebrity role models** — tend to promote the 'glamour' of having a career and of returning to work immediately after giving birth, making use of child-minding services

In the second part of *Case study 1*, reference was made to a recent increase in the UK's population. Currently, the growth is explained by rising immigration rather than by natural increase. However, given the nature of many of the new immigrants — essentially young people drawn from societies where traditional attitudes about large families persist — it seems likely that Britain's birth rate may soon be lifted out of the doldrums. The steady rise in life expectancy (looked at more closely in *Case study 33*) continues for most gender/age groups in the UK. Much of the reduction in British mortality rates is due to improvement in the survival rates among people afflicted by the four main killer diseases. During the 1990s:

- deaths from coronary heart disease fell by 36% among men and 40% among women
- lung cancer deaths fell by 28%
- the incidence of breast cancer in women fell by 24%
- the number of stroke victims fell by 30%

A government prone to spin would claim that this is the outcome of a 'massive' increase in spending on the NHS. In 1990, the UK was spending 6.2% of GDP on healthcare; in 2010 the figure reached 8.3%. In contrast, France spent about 9.3% per annum over the same period. The lower mortality figures are more likely to be the outcome of education, changing lifestyles and earlier diagnosis.

With fewer births and deaths, the overall population becomes still 'greyer'. This is reflected in the fact that 15% of the population is now aged 65 and over. The slight decline in the **dependency ratio** from 0.62 in 1971 to 0.55 in 2001 and its rise to 0.66 in 2010 conceals the fundamental shift in the character of dependency (Figure 2.12). The percentage of children in the population has slipped to 20%. Projections suggest that by 2014 there will be more over-65s than under-16s.

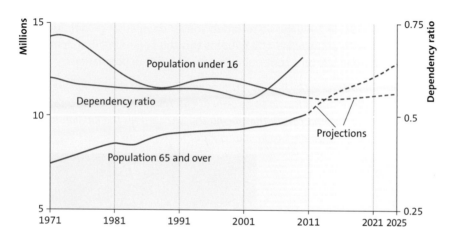

Figure 2.12
Changing dependency in the UK, 1971–2025

If we believe that the DTM is in the process of developing a fifth stage, then the following demographic characteristics, shown in this case study, could be among the hallmarks:
- *the lowering of an already low birth rate*
- *first-time mothers becoming older*
- *the rise of childless marriages and partnerships*
- *a change in traditional lifestyles, such as more same-sex relationships*
- *a rise in the death rate, despite health improvements as the population increases in 'greyness'*
- *changing dependency — the shift from children to pensioners*
- *the potential decline in population offset by immigration*
- *immigration increasing ethnicity*

One question arising from this case study is: do the explanations of persistently low birth rates and declining death rates in the UK apply to other countries? For example, how is it that Catholic countries such as Italy and Spain (see Case studies 28 and 31) *now have the lowest birth rates in Europe? How is it that the UK mortality rate has been falling more quickly than that of other western European countries?*

Russia is facing a demographic crisis as the live birth rate plummets and the death rate is raised through excessive consumption of tobacco and alcohol. Today, less than one-third of recorded pregnancies produce live births. The plunge in the live birth rate has been brought about largely by poverty (Figure 2.13), by illnesses (e.g. anaemia, heart problems and malnutrition) among pregnant women and by abortion being used as a form of birth control. Adding to the agony is the high rate of infant mortality. Recently released figures indicate that 30% of all Russian newborn babies die of infections. A large number of babies born with correctable heart defects are not being operated on. Healthcare has been one of the worst victims of Russia's transition from a socialist to a market economy. The system has crumbled. At the same time, cigarettes and vodka have lowered life expectancy to among the shortest in the world. Average male life expectancy is now only 62 years — 10 years less than the mean expectation in western Europe and less than it was in communist times.

As a consequence of the decline in live births and the increases in infant and adult mortality rates, the total population of Russia has fallen by over 6 million from a peak of 149 million in 1991. In 2009 Russia recorded annual population growth for the first time in 15 years, with an increase of 23,300! Is this the beginning of a demographic recovery or is it a small blip on a continuing downward shift in population?

The immediate question raised by this case study is whether or not the Russian predicament is a one-off situation. If not, is Russia to be seen as entering Stage 5 in the DTM, but one that is perhaps rather different from that facing the UK? Or does the Russian experience simply represent a regression back to an earlier stage in the model?

Figure 2.13
A poor family in Moscow

Ullstein Konzept und Bild/Still Pictures

Using case studies 4

Look back at the population pyramids in Figure 2.9. Population pyramid A shows the Russian Republic and pyramid B shows the UK.

Question

The UK and the Russian Republic may be two of a small number of countries pioneering what is now being seen as Stage 5 in the DTM. Using *Case studies 7* and *8*, compare their present demographic characteristics. What conclusions do you reach?

Guidance

- The pyramids in Figure 2.9 suggest that the population trends have been more consistent in the UK; volatile change is evident in the Russian pyramid.
- The UK population is rising slowly; that of Russia is falling.
- The increase in the UK can be explained by immigration; possibly emigration is playing its part in Russian decline.
- Birth rates are declining in both countries.
- The death rate is falling slowly in the UK, but rising in Russia; the rise in infant mortality is particularly marked.

Having completed the above exercise, you might conclude that there are some major differences between the UK and Russia. It is difficult to believe that both countries are passing through the same demographic transition stage. Maybe we have to recognise that there are two variants of Stage 5. In one of them, there is recovery from a period of stagnant population. In the other, there is a sustained decline in population that eventually levels out.

Part 3

Population distribution

The global pattern

Awareness of the difference between distribution and density is vital to understanding this component of population geography. **Distribution** is where people are located in an area. Population distribution is most simply shown on a map by representing each person or group of people by a dot or symbol. This shows us where people are and gives an impression of how the numbers of people vary from place to place. One of the most useful indications of population distribution, however, is achieved when population numbers are related to the space they occupy. This is **density** — the number of people per unit area (km^2 or hectare).

The most crucial fact of population geography is that people are not evenly distributed across the face of the Earth. This can be seen at a range of spatial scales from global to local. In 2010, the world's population was an estimated 6.8 billion, occupying the world's land area, including Antarctica, of 150 million km^2. This

Figure 3.1
The global pattern of population densities, 2006

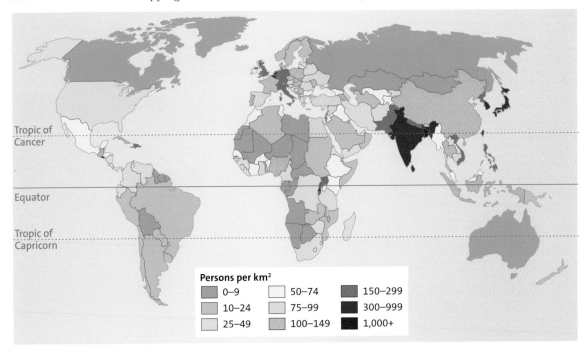

Persons per km²
- 0–9
- 10–24
- 25–49
- 50–74
- 75–99
- 100–149
- 150–299
- 300–999
- 1,000+

Tropic of Cancer

Equator

Tropic of Capricorn

gives a mean population density figure of 45 persons per km^2. Over two-thirds of the global population live in Europe and Asia, but these two continents account for little more than one-third of the total land area.

While continental population densities range from 3.6 persons per km^2 in Australasia to 87.0 persons per km^2 in Asia, looking more closely at a population distribution map reveals far greater extremes (Figure 3.1). There are considerable variations at a national level. In Asia, for example, density ranges from 2,700 persons per km^2 in Singapore to a mere two persons per km^2 in Mongolia.

5 **Using case studies**

Summarising the essential features of a distribution map

The aim here is to improve your ability to describe the essential features of any distribution map. This is a task frequently required in examinations. The global distribution of population (Figure 3.1) is a good map to focus on. Of all the global maps showing different aspects of population (change, birth rates, death rates, life expectancy, etc.), this is perhaps the most helpful in revealing the basic demographic scenario of the world.

In general terms, the description strategy should be to go for the general rather than the particular. In other words, try to identify the really outstanding features. A useful tip here is to focus first on the 'extreme' areas: namely, those showing the highest and lowest values. Deal with each of these in turn by means of a few broad brush-strokes (i.e. bullet points). Pointing out exceptions to the generalisations you have made should be kept to a minimum. Finish by focusing on the middle-value areas.

How does this work out in practice? Let us look at Figure 3.1 and tackle the question that follows.

Question

Identify the main features of the global distribution of population.

Guidance

- High population densities (say, over 75 people per km²) occur in (1) Asia (the Indian subcontinent, China, Korea, Japan and much of the southeast), (2) Europe and (3) the Caribbean.
- Population densities at or below the global average of 45 people per km^2 prevail (1) in North and South America, (2) over much of Africa and the Middle East, (3) across northern Eurasia and (4) in Australasia.
- A significant feature of Africa is the extent to which population densities vary quite abruptly from country to country.
- Remember that the question has not asked us to explain, just to identify (i.e. describe).

You might compare the distribution shown in Figure 3.1 with that in Figure 2.4. A combination of a high density of population and a high rate of population growth is going to indicate possible population pressure points. If they are not already, there is a strong likelihood that such areas could become areas of **overpopulation**. It is interesting to note that Africa, with its generally low densities, shows some of the highest rates of population growth.

Factors affecting distribution and density

The two case studies that follow illustrate that, as the spatial focus is narrowed, yet more density contrasts become evident, between regions and, most strikingly, between urban and rural areas.

The number of people in a given area, and their distribution within it, are the outcome of a complex interaction of factors, as shown in Figure 3.2. Present population should be thought of as the product of population trends operating over long periods of time. These population trends are not always upward. Populations can and do stagnate; they can even decline. In short, **population change** does not always mean population growth. As will be seen in the next section, change is the result of the overall and constantly shifting balance between births, deaths and migration.

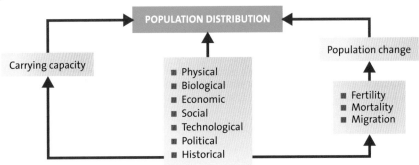

Figure 3.2
Factors affecting the distribution of population

Figure 3.2 shows that population distribution is also influenced by another key factor, **carrying capacity**: namely, the maximum number of people that can be supported by the resources and technology of a given area. However, there is a wide range of factors affecting the key influences on population distribution. These factors might be thought of as both creating opportunities and setting limitations (i.e. constraining).

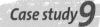 **Case study 9** POPULATION DISTRIBUTION IN ETHIOPIA

Influential factors

Ethiopia's population size and rate of growth are among the highest in Africa. According to the 2007 census, the population was 73 million, making it the second most populous sub-Saharan African country after Nigeria. With an annual growth rate estimated at 2.6%, Ethiopia's population will approach 110 million before 2020 — double what it was at the time of the 1994 census. This rapid population growth, the result of high fertility and a rapid lowering of the death rate, represents a serious obstacle to Ethiopia ever achieving sustainable development.

Over 80% of Ethiopians live in rural areas and a significant percentage of these people are pastoral nomads, which makes counting the population both difficult and inaccurate. With per capita GDP at only a little over $100, Ethiopia is one of the poorest countries in the world.

Looking at the distribution of population in terms of density, the national average is 72 persons per km². At a regional level, densities range from less than 30 to over 80 persons per km² (Figure 3.3).

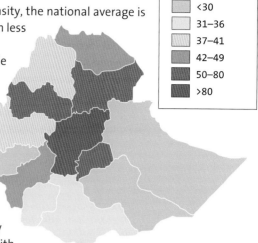

The distribution pattern is strongly conditioned by altitude and relief. Physically, Ethiopia consists of two great plateaus separated by part of the Great Rift Valley (Figure 3.4).

The Ethiopian Plateau to the west of the Rift Valley is the most fertile and populated part of the country. The highest densities of population centre on the capital city, Addis Ababa (population 3.4 million). East of the Rift Valley is the Somali Plateau, which rises to over 4,250 m in the Bale Mountains before sloping gently eastward to the Ogaden Plateau. The Rift Valley separating the two plateaux is a long, narrow cleft dotted with lakes that broadens to the north to form the Danakil Depression, an extensive desert.

Figure 3.3 Ethiopia: population distribution, 2005

The large range in altitude, from less than 500 m to over 4,000 m, has a considerable impact on climate and it is climate that ultimately affects the distribution of population. About 15% of the population live in areas above 2,400 m, around 75% live in the zone between 1,500 and 2,400 m, and only about 10% live below 1,500 m — despite the fact that over half of Ethiopia's territory falls in this category. Locations above 3,000 m and below 1,500 m are sparsely populated. Above 3,000 m the terrain

Figure 3.4 Ethiopia: relief and drainage

is rugged and temperatures are low. These factors limit agriculture. Below 1,500 m (apart from in the west and southwest) there are high temperatures, low rainfall and recurrent drought.

Strong though the influence of the physical environment is on the distribution of population, other factors have been, and are still, at work. A number of non-physical factors indicated by Figure 3.2 are outlined below:

- **Biological factors:** rates of population growth are high in the most populated and more prosperous regions. Malaria and other chronic illnesses are common in the lower parts of the country and act as a deterrent to settlement.
- **Economic factors:** the economy is based on agriculture; over 75% of the population is dependent

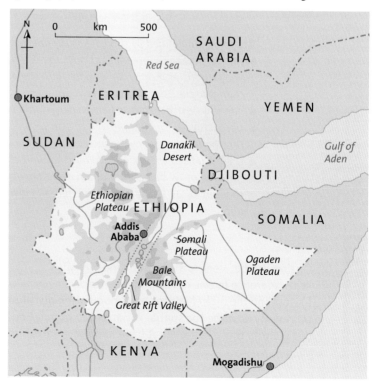

Figure 3.5
Ethiopia: the level of need for emergency food assistance

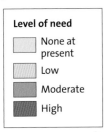

Level of need
None at present
Low
Moderate
High

TIGRAY

AFAR

AMHARA

BENSHANGUL

GAMBELA

OROMIA

SOMALI

SNNPR

directly on farming and livestock rearing. Developing other means of livelihood might help to raise the national carrying capacity, but the opportunities are limited (tourism is one possibility). However, they are unlikely to make much impact on the overall distribution of population.

■ **Social factors:** massive rural–urban migration to Addis Ababa helps to concentrate the population on the Ethiopian Plateau.

■ **Technological factors:** because Ethiopia is such a poor country, the available technology can do little to open up the more remote areas and to improve the agricultural productivity of more marginal environments.

■ **Political factors:** a longstanding border dispute with Eritrea (formerly part of Ethiopia) means that money that could be used to deal with the demographic situation is spent instead on the armed forces. Inhabitants have been forced to move away from the dangerous frontier zone.

The net effect of these and other factors is to preserve and accentuate the historical pattern of an unevenly distributed population. As a consequence, carrying capacities in the favoured areas are being pushed to their limits. Indeed, malnutrition and imminent starvation indicate that these limits have already been exceeded in some places (Figure 3.5). Solutions to this desperate situation are hard to see, except possibly that with an HIV prevalence rate estimated at around 10% (one of the highest in Africa) and low use of contraceptives, rising deaths from HIV/AIDS might curb the rate of population growth and perhaps even change the distribution pattern.

The points underlined by this case study are that:

■ *the physical environment continues to exert a powerful influence on the distribution of population, especially in a poor developing country, because possible means of changing the established pattern are severely limited*

■ *non-physical factors also help to perpetuate the pattern inherited from the past*

■ *uncontrolled population growth can result in carrying capacities being exceeded, a threshold indicated by the increased incidence of poverty, malnutrition and starvation*

■ *there is a need to take into account the vertical dimension in a subject that is generally thought of as being only about the horizontal. Generally speaking, high altitude is thought of as being a negative factor. However, in Ethiopia the situation is rather different.*

6 **Question**

Using case studies

(a) Distinguish between 'distribution' and 'density'.

(b) With reference to one developing country:

(i) outline the main features of its population distribution

(ii) briefly describe the spatial variations in population density

(iii) identify those factors that have had the greatest impact on population densities

Contemporary Case Studies

(a) Distribution is about the actual location of people within an area. Density is the number of people per unit of land area. A located dot map shows distribution, whereas density is shown on a choropleth map.

(b) Ethiopia is used as the example.

(i) The distribution is predominantly rural, although 1 in 20 people live in the capital city, Addis Ababa; most of the population lives in the western half of the country; 75% of the population live at an altitude between 1,500 and 2,400 m — 15% live above, and 10% live below that range.

(ii) The mean population density is 72 persons per km^2. However, at a regional level, densities range from less than 30 to more than 80 persons per km^2. The highest population densities occur in a belt running roughly north–south along the highlands, immediately to the west of the Rift Valley. Much of the eastern part of the country is sparsely populated.

(iii) Altitude is a major influence; population densities are lowest on the highest and lowest ground. Altitude tends to makes its impact through the medium of other factors, particularly climate, carrying capacity and economic opportunities.

Other factors include:
- disease
- inertia — the pattern inherited from the past and persistent poverty
- war — particularly near the Eritrean border
- urbanisation — particularly rural–urban migration to Addis Ababa

Limitations of distribution maps

Having read and digested the case study of Ethiopia, it could be tempting to think that the grip of the physical environment on population distribution is going to be much less in developed countries. After all, their governments have the resources, organisation and technology to intervene and manage. Japan, for example, is one of the most densely populated and economically developed countries in the world. Surely the Japanese have the upper hand when it comes to dealing with the physical environment? The following case study may persuade you to think otherwise.

A WARNING ABOUT DISTRIBUTION MAPS FROM JAPAN

Case study 10

With a mean population density of 336 persons per km^2, Japan is one of the world's most densely populated countries. When we look at the distribution at a prefectural level (a prefecture is the Japanese equivalent of a UK county), densities range from 68 persons per km^2 in the northernmost and rather inhospitable island of Hokkaido to over 6,000 persons per km^2 in Tokyo Prefecture, which contains a large part of the capital's metropolitan region.

At face value, the map produced by plotting the density figures for the 47 prefectures (Figure 3.6) would seem to give an accurate picture of the distribution of population in Japan. However, if you take a look at any physical map of Japan, you will see that it is dominated by mountainous terrain. Although rarely rising above 1,500 m, this terrain is

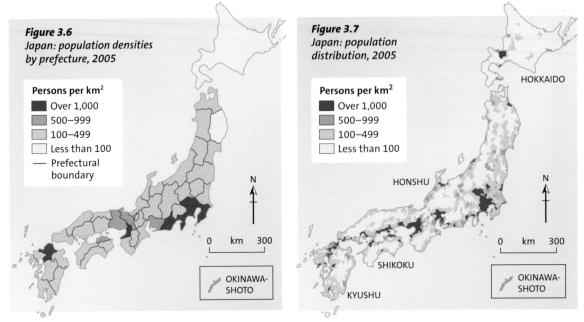

Figure 3.6
Japan: population densities by prefecture, 2005

Persons per km²
- ■ Over 1,000
- ■ 500–999
- ■ 100–499
- □ Less than 100
- — Prefectural boundary

N

0   km   300

OKINAWA-SHOTO

Figure 3.7
Japan: population distribution, 2005

Persons per km²
- ■ Over 1,000
- ■ 500–999
- ■ 100–499
- □ Less than 100

HOKKAIDO

HONSHU

N

0   km   300

SHIKOKU

KYUSHU

OKINAWA-SHOTO

deeply dissected by V-shaped valleys separated by narrow ridges. In terms of settlement and development, this upland country has been a persistent handicap, particularly to transport. Even today, it remains largely unpopulated and unused.

In stark contrast to the all-prevailing uplands are the alluvial lowlands, which account for just one-eighth of the land area. These lowlands are highly fragmented, small and largely coastal. It is here that most Japanese live and work, mainly in towns and cities. The actual population densities are far higher than the prefectural means shown in Figure 3.6. So, in a sense, Figure 3.6 may be accused of concealing reality. Figure 3.7 gives a more accurate picture and draws our attention to the following key features:

- ■ the relative emptiness of much of inland Japan
- ■ the attraction of population to the coastal lowlands
- ■ the corridor of very high population densities running along the south coast of Honshu from Tokyo westwards and perhaps reaching as far as the northern part of the island of Kyushu. This corridor is occupied by large urban agglomerations that are tending to grow together to form one huge, continuous urban complex (Tokaido megalopolis).

With population densities at levels well over 1,000 persons per km², it is intriguing to ask why the Japanese have not used their immense wealth and great technological competence, not just to increase the country's **carrying capacity**, but also to make its population distribution more even. Is it because they recognise that the physical environment is supreme and that in the long run it is better to work with it rather than against it? That may well be the case, but the persistence of the coastal pattern is encouraged by historical factors (the momentum of the past) as well as by economic considerations, such as the importance of the sea to Japan's food supply and economy (importing raw materials and energy, and exporting finished goods).

This case study exemplifies a number of points:
- ■ The physical environment can be an important influence on the distribution of population even in developed countries.

■ *Strong contrasts in population density can exist within highly populated countries.*
■ *Choropleth maps can be deceptive of as a means of showing the actual distribution of population.*

Changing distributions

The population distribution map is a snapshot in time. It freezes a frame in what is in fact a continuous sequence of change. At an international level, the major forces driving that change are national variations in population growth rates — i.e. natural change. When we zoom in and take a closer look at the changing distribution within a particular country, regional and local variations in natural growth rates often become much less significant. Instead net migration becomes rather more influential.

SIGNIFICANT SHIFTS IN BRITAIN

Case study **11**

Between the censuses of 1951 and 2001, the total population of the UK rose by 17% from 50.2 million to 58.8 million. Much of this increase was due to natural change. But migration also played a part, not just through the arrival of immigrants but also via their impact on birth rates. In this case study, we will look at how the distribution of population in Great Britain changed during three sample intercensal periods: 1951–61, 1971–81 and 1991–2001. You should note that the scale of shadings used in each of

Figure 3.8
The changing distribution of population in Britain (percentage change)

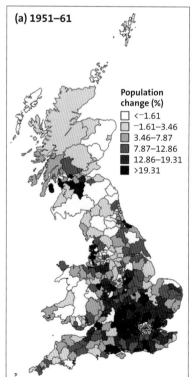

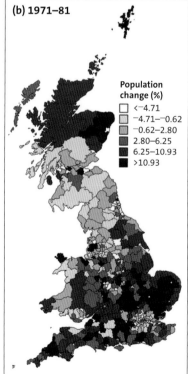

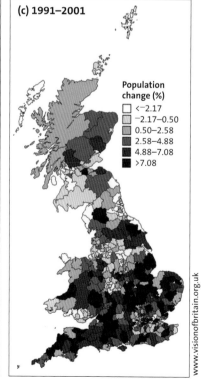

www.visionofbritain.org.uk

the three maps in Figure 3.8 is slightly different. This does not impede the general portrayal of shifts in the overall distribution of population.

In the 1950s the UK population grew by 5.0%. This coincided with the rising limb of the postwar 'baby boom'. It also resulted from modest net in-migration — a large influx of New Commonwealth immigrants was balanced by substantial emigration to the USA, Canada, Australia and New Zealand. Figure 3.8 (a) shows well the beginnings of what became known as the 'drift to the south'. Notice the negative and close to zero growth rates over much of Scotland, Wales and northern England. In contrast, high rates of growth are very evident over much of England, particularly the southeast. The overall pattern suggests that people on the move from remoter and more peripheral parts of Britain are being drawn by the job opportunities and bright lights of Greater London and the other conurbations. Immigrants, particularly from the Caribbean, also helped to swell the populations of London, Birmingham and other major cities. They had been specifically recruited to help the persistent shortage of labour.

In the 1970s, the UK population grew by only 0.8%. This was the result of substantial net out-migration and a birth rate at or close to an all-time low. The increasing concentration of population in the Midlands and southern England is still very much in evidence. But three new features are apparent in Figure 3.8(b). One is a loss of population from the **core** areas of all the conurbations — Greater London, Birmingham, Manchester, Merseyside, Sheffield and Newcastle-upon-Tyne. The second is a gain in population in the remote rural areas of southwest England, Wales and northern Scotland. A third gain in population is also apparent in the accessible rural areas of England as the commuter catchments of the major cities were extended outwards. All three gains are indicative of **counter-urbanisation** — a degree of dissatisfaction with big-city living.

In the 1990s, the UK population grew by 2.9%. This was due to substantial net in-migration, much from the EU, especially new member states in eastern Europe. Although the birth rate was higher than in the 1970s, it fell slightly during the decade. The main population growth remains focused on England and on accessible rural areas, particularly in northern England, the east Midlands, East Anglia, and southern and southwest England. Much of Scotland either lost population or showed minimal growth.

The main outcome of 50 years of population change in Great Britain has been that southern England has become even more marked as the country's demographic centre of gravity. It is the small towns and accessible countryside rather than the major conurbations that have attracted much of this shift in population. Clearly, counter-urbanisation has made its mark. Remember, however, that the variations in 10-year population growth rates shown in Figure 3.8 are the product of both natural change and migration. For the former, variations in the age structures of populations have had an impact via birth and death rates. The distinction between 'young' and 'old' populations is significant. Migration flows have either accentuated or ameliorated the pattern of natural change.

This case study illustrates the dynamic nature of population distribution. It is constantly changing in response to a range of factors other than just rates of natural and migration change. Significant economic and social factors add to the dynamism, as can political intervention.

Reference to migration leads us neatly into Parts 4 and 5, but look particularly at *Case studies 14, 20 and 33* for more information about processes and flows that have contributed to changing distribution of population in Britain.

People on the move

Types of movement

In studies of population, the term 'movement' (or **mobility**) has two components — migration and circulation. Both involve people shifting locations. **Migration** refers to those moves involving a change of residence that lasts for at least 1 year. **Circulation** involves moves that are shorter term; the changes in location are in a sense temporary. Shopping, commuting, tourism, pastoral nomadism and shifting cultivation are examples.

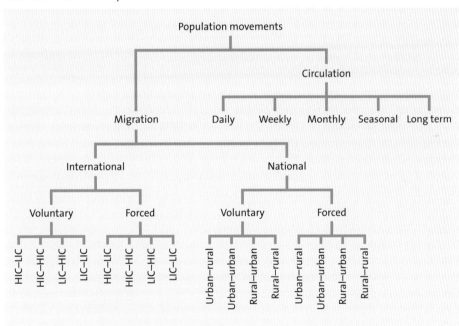

Figure 4.1
A classification of population movements

It will only take you a little time to come up with a list of population movements. No doubt your examples will be drawn from different parts of the world. It is also likely that they will differ in terms of scale (i.e. numbers of people and distances involved), when they occurred and for what reasons. Figure 4.1 suggests one possible classification that helps us to put these and many more examples into a single framework. Migration and circulation represent the first-order subdivision. Circulation is then subdivided on the basis of its frequency. Equally, it could be broken down on the basis of its purpose, as suggested by the examples in the previous paragraph. The classification of migration is rather more complicated.

There are many possible criteria:

- distance
- direction
- volume
- cause
- motivation
- the nature of the decision-making process

The distinction between national and international migration is important because it acknowledges the significant impact that state boundaries can have on population movements. At the next level, the distinction between forced and voluntary migration is thought to be of great significance. The final tier in the classification emphasises the significance of direction: for instance, between urban and rural areas within a country and between developing countries and developed countries at a global or international level.

Geographical studies of migration have two main focuses: motives and impacts. There are at least two sides to each of these focuses. In the case of the motives (examined in this part of the book), there are both **push and pull factors** to be considered. Similarly, as you will see in Part 5, the impacts of migration on the **source area** are very different from those on the **host area** or destination.

Case study 12 — GLOBALISATION AND ECONOMIC MIGRATION

Globally, migration has doubled over the past two decades and seems likely to continue accelerating, affecting every country in the world. The number of migrants worldwide increased from 84 million in 1975 to 191 million in 2009. Current predictions say that the rise will soon begin to level off, but nonetheless the number of migrants is expected to reach 230 million by 2050.

About 23 million people emigrate from developing countries to developed countries each year. These economic migrants now account for two-thirds of population growth in the North. The migration dream is being encouraged through imported television programmes, advertising of imported luxury goods and talk of high wages, which persuade citizens of developing countries that a much more comfortable lifestyle is to be found in Europe, the USA and Australasia. However, a major driving force behind the South–North movement of people comes from the prevailing demographic situations. On the one hand, the rapid population growth in the South increases the pressure to emigrate to the North in search of work. At the same time, labour, particularly unskilled labour, is in increasingly short supply in the 'greying' populations of the North. In short, a one-way labour migration is helping to solve both problems.

Three other factors have played a part in conditioning the pattern of global migration (Figure 4.2) and raising the volume of migration:

- Some political barriers have been removed: for example, in the break-up of the Soviet Union and China's relaxation of travel restrictions.
- Cheaper long-distance transport facilitates international migration.
- The communications revolution (television, the internet, mobile phones) has resulted in a much greater awareness of opportunities overseas.

However, not all of today's international migrants are searching for work.

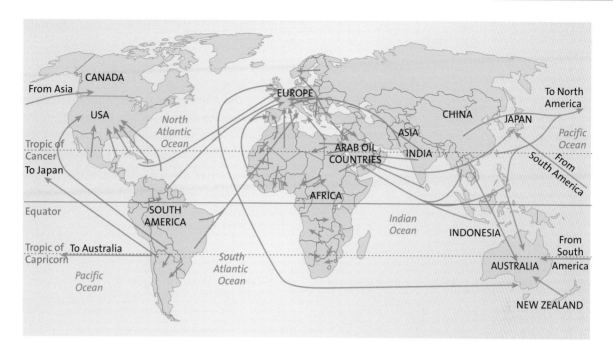

Figure 4.2
The main global migrations of the last 25 years

Since the end of the Second World War in 1945, there have been over 150 other wars, mostly in developing countries. These wars, together with persecution policies based on **ethnicity** (as in Bosnia, Serbia, Kosovo, Iraq and Afghanistan), have generated great streams of **refugees**. People have abandoned their homes in fear. It is estimated that there are some 40 million refugees in the world today. Environmental refugees are growing in number, but they are driven by yet another basic human need — food (see *Case study 15*).

For centuries, Europe was the continent of emigration. That has now changed. Asia has become the number-one source of emigrants, the majority coming from India, Bangladesh, Pakistan, China, Vietnam and the Philippines. At the same time, Europe has become the immigrant honeypot. Here, 56 million people (nearly 8% of the population) are living outside their country of birth. Although the absolute numbers are smaller in North America and Oceania (Australasia), the percentage figures are higher, at 13% and 19% respectively. In contrast, the populations of many developing countries are relatively unaffected by immigration, but are more so by emigration. Immigrants account for only about 1% of the population of Asia and Latin America, and 2% of that of Africa.

The present global migration from developing countries to developed countries can be seen not just as an acceleration, but also as a continuation of migration flows that have waxed and waned over the centuries; flows that have changed the face of the world. Figure 4.2 identifies some of the main global migrations of the last 25 years.

Government policies can have a major impact on international migration. They can either make it happen or prevent it. The latter is more common, as more and more countries are either closing their borders or severely restricting in-flow. Twenty-five years ago, only 6% of countries had policies to curb immigration; now 40% of countries do. There are only five countries in the world today that have official policies to encourage permanent in-migration. They are the USA, Canada, Australia, New Zealand and Israel.

One of the directors of the International Organisation for Migration said recently: 'Migration is a reality; the issue is how to make it work to benefit both migrants and the receiving populations.' Unfortunately, there are other issues that also need to be sorted out, such as bogus asylum seeking (see *Case study 35*), illegal trafficking in people (*Case study 13*) and tighter border controls that threaten the human right to freedom of movement.

The rising tide of global migration is mainly the outcome of the widening gap between the rich and poor nations of the world. Differences in living standards, welfare and opportunities for betterment create a powerful alignment of pull and push factors. They are the very stuff on which migration feeds.

7

Using case studies

Question

Summarise the main global migrations of the last 25 years, as shown in Figure 4.2.

Guidance

- 'Summarise' is the vital command word. It means identify the main flows. Do not indulge in a tedious flow-by-flow description. In an exam, the mark allocation, together with the lined space allocation, should indicate the degree of generalisation required by the examiner.
- Focus first on the main destinations of intercontinental migration, as indicated by the convergence of the flow lines. These are the USA, Europe, the Middle East and Australia.
- If there is space, outline the main source regions for each of these destinations.
- Africa deserves a quick mention because there was a fair amount migration within the continent; to a lesser extent, so does South America.
- Remember, you have only been asked to summarise, not to explain, what you see on the map.
- In today's world on the move, there are undercurrents of illegal migration across national frontiers.

While many would argue that freedom to roam should be a basic human right, in reality national frontiers often act as migration barriers (see *Case study 35*). This applies particularly to would-be immigrants. For reasons of protection and the national interest, inward flows are increasingly controlled by quotas and visas. Inevitably, where legal barriers are placed in the way of people, illegal ways will be found, often by the criminal class, to get round those barriers. Thus it is in today's world on the move that there are undercurrents of illegal migration across national frontiers. The pathways of these illegal migration flows may be generally known, but no one knows the exact number of people moving along them.

Case study **13** **ILLEGAL IMMIGRATION IN THE EU**

Figure 4.3 gives a fairly recent picture of the distribution of net migration within the EU. The new member (A8) countries in eastern Europe show negative migration balances (see Figure 5.7 on page 63). Many young adults have left in search of work and better rates of pay (see *Case study 19*). The Mediterranean countries show high net migration

Figure 4.3
Net migration in the EU, 2000–04

B   Belgium
CZ Czech Republic
L   Luxembourg
N   Netherlands
S   Slovenia
SL Slovakia

Migrants per 1,000 population
- 31+
- 12–30
- 11–20
- 1–10
- −10–0

gains. This reflects the north to south movement of 'sun seekers' (especially retired people — see *Case study 21*) plus illegal immigration from Africa (Figure 4.4). Africa is only one of several sources of people entering the EU illegally: that is, without visas or work permits. The 'new' eastern border of the EU with Moldova, the Ukraine, Belarus and Russia is also widely recognised as being 'highly porous' and allowing the illegal entry of thousands of migrants from the Middle East and southern and southeast Asia.

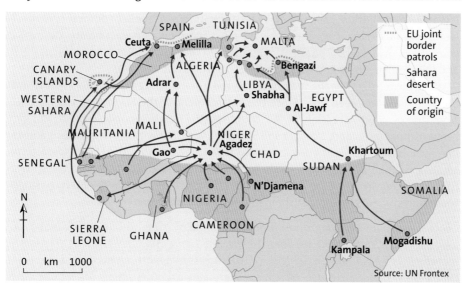

Figure 4.4
Illegal immigrant routes into the EU from Africa

Source: UN Frontex

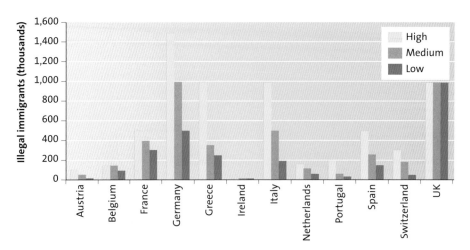

Figure 4.5
Estimates of illegal immigrants in 12 EU countries, 2003

For obvious reasons, there are no accurate figures for illegal immigration into the EU. The only accurate data relate to the number of illegal immigrants actually apprehended at border crossings. But these are merely the tip of the iceberg. Estimates of the number of illegal immigrants living in the EU made in 2003, just before the enlargement of the EU, ranged between 2.6 and 6.4 million.

Figure 4.5 shows three levels of estimate for 12 member countries. These clearly show Germany and the UK as the most popular illegal immigrant destinations. The fact that three estimates for the UK are all the same suggests the availability of some reliable data. However, an official report in 2006 estimated the number at 500,000 — due, in part, to disinformation about job opportunities and access to social services. It may be that migrants have some familiarity with the English language and as a result feel less daunted by moving to an English-speaking country. It is also clear that there is much deception, in that many illegal immigrants are led to believe that they are moving to a country where they can escape the pressures and disadvantages of their homeland.

Deception is the tool of the trade for those who, for considerable sums of money, are wiling to smuggle people into the UK and other EU member states. There is quite clearly an underworld of human traffickers. **Human trafficking** is the modern form of slavery. It is about exploitation, which can take the form of forced labour or domestic service, as well as sexual exploitation. The last is the most common and overwhelmingly involves young women and girls. The arm of human trafficking to the UK reaches as far as Russia and the Ukraine, China and the Philippines, as well as west Africa.

While eastern Europeans may have figured among smuggled workers in the past, there is much less need now. Since the enlargement of the EU in May 2004, citizens from the so-called 'Accession' or 'A8' countries (the Czech Republic, Estonia, Hungary, Latvia, Lithuania, Poland, Slovakia and Slovenia), together with those from Bulgaria and Romania, have enjoyed the legal right to travel freely within the EU. In some of the established member countries, such as the UK, they have also been allowed to work (see *Case study 19*). However, illegal immigrants continue to enter the EU by posing as tourists or students before vanishing into the black economy.

There is no doubt that migration is one of the major issues in our rapidly globalising world. There is an in-built tension. On the one hand, most believe that freedom of movement is a basic human right, along with access to opportunities of self-betterment. On the other hand, it is widely recognised that many countries simply cannot open their doors indefinitely and indiscriminately to all would-be immigrants. This tension creates circumstances that positively encourage illegal immigration and human trafficking. (See Case study 35.)

Migration balances

For most areas or countries, migration is a two-way traffic. There will be people arriving at the same time as people are leaving. The balance between the two opposing migration flows is critical. This is referred to as **net migration** (Figure 4.3). **Positive net migration** means that there are more incomers than out-goers. The reverse situation gives rise to **negative net migration**. It may seem strange that at any one time there will be people moving along the same migration route but in opposite directions. This is well illustrated by what are probably the world's leading migration flows: namely, those between rural and urban areas.

MIGRATIONS BETWEEN TOWN AND COUNTRY IN THE UK

Case study 14

Of all the global migration flows, by far the greatest have been, and still are, those between rural and urban areas. Furthermore, most of these flows are domestic: in other words, the flows are much greater within rather than across national frontiers. In short, we are dealing here with various forms of domestic or national migration. By far the most widespread of the migration flows is rural-to-urban migration, a vital process in the early stages of urbanisation. However, progress along the urbanisation curve leads to the start of a new set of processes and associated migration flows (Figure 4.6). They are:

- **urbanisation** — rural–urban migration and centralisation continue
- **suburbanisation** — intra-urban and urban–rural migrations
- **counter-urbanisation** — urban–urban and urban–rural migrations
- **reurbanisation** — urban–urban and rural–urban migrations

Although the UK is one of the most highly urbanised countries in the world, all four processes and their associated migrations are at work. It is these flows that have played a part in the changing distribution of population in postwar Britain (see *Case study 11*).

With such a complex of migration flows, it is extraordinarily difficult, indeed impossible, to make any meaningful comments about migration balances. The only overall measure would be the percentage of the population living in an urban environment. Any increase might suggest an urban net migration gain. Visual evidence of the flows and their

Figure 4.6
Rural–urban spatial processes and migration flows

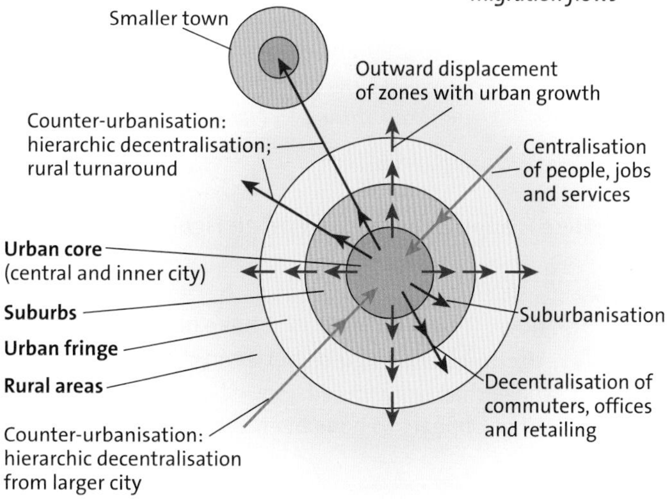

Smaller town

Outward displacement of zones with urban growth

Counter-urbanisation: hierarchic decentralisation; rural turnaround

Centralisation of people, jobs and services

Urban core (central and inner city)

Suburbs

Urban fringe

Rural areas

Suburbanisation

Decentralisation of commuters, offices and retailing

Counter-urbanisation: hierarchic decentralisation from larger city migration balances might take the form of old and underused housing in areas of net loss and new housing in areas of net gain.

Motives for migrating

It is no exaggeration to say that migration is one of today's major global issues (see Part 8). The growing volume of international migration draws our attention to the persistence of unbearable problems in the main source regions. Equally, these same flows are creating some serious problems in the more popular destination regions.

Figure 4.7
The 'push–pull'
mechanism

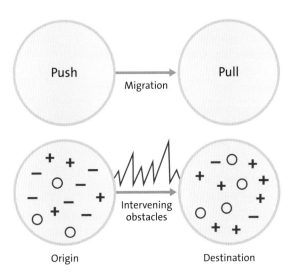

+ Positive factors
− Negative factors
○ Neutral factors

It is widely held that most migration is the outcome of two sets of forces (Figure 4.7). **Push forces** work in the migrant's current location. They can range from unemployment to persecution; from natural disasters and famine to poverty and war. The **pull forces** are those that attract the migrant to a particular location. They are mainly the mirror images of push factors and therefore include attractions such as job opportunities, tolerance, personal safety, freedom, good housing and welfare services. Intervening obstacles, both perceived and real, need to be overcome before migration takes place. They include international frontiers and the often considerable costs of moving. Facilitating factors that encourage migration include culture, language, the media, perception and ease of access to visas and permits.

The decision to move will rarely be made on the basis of a single push or pull factor. Rather, the decision will be based on an appraisal of a range of attributes, both in the place of origin and at the potential destination. Each person perceives these attributes differently, depending on personal characteristics such as age, gender, socioeconomic class, occupation and education. Some of the attributes at the present location will be regarded positively, persuading the person to stay put. Some will be seen negatively, encouraging migration. Other attributes will be perceived neutrally and thus have no bearing on the decision making. The same threefold classification applies to the potential destination area, except that here the positive factors encourage, and negative factors discourage, migration.

8 Here is another exercise based on Figure 4.2.

Using case studies

Question

The flows shown on Figure 4.2 have not been classified.
(a) Attempt to distinguish between those migrations that were essentially voluntary and those that were forced.
(b) What appears to have been the most powerful push factor?

Over the last 150 years, the concept of push and pull has been refined, and a range of other ideas has been applied to the study of migration. They are summarised in Table 4.1.

Name	Date	Key ideas
Ravenstein's laws of migration	1875–89	Most migrants travel short distances and with increasing distance the numbers of migrants decrease. Migration occurs in a series of waves or steps. Each significant migration stream produces, to a degree, a counterstream. Urban dwellers are less migratory than rural dwellers. The major causes of migration are economic.
Stouffer's theory of intervening opportunities	1940	The volume of migration between two places is related not so much to distance and population size, but to perceived opportunities that exist in those two places and between them.
Zipf's inverse distance law	1949	The volume of migration is inversely proportional to the distance travelled by migrants, and directly proportional to the populations of the source and destination.
The gravity model	1960s	This simple formula expresses Zipf's two relationships.
The Lee model	1966	This revises the simple 'push–pull' model in two ways: It introduces the idea of 'intervening obstacles' that need to be overcome before migration takes place. Source and destination are seen as possessing a range of attributes; each would-be migrant perceives these attributes differently, depending on personal characteristics, such as age, gender and marital status.
The Todaro model	1971	This stresses that potential migrants weigh up both the costs and benefits of moving before taking any action; migrants act in economic self-interest.
The Stark model	1989	This extends the Todaro model by arguing that there is more to migration than the optimising behaviour of migrants; risk spreading in families is one such factor.
Marxist theory	Fashionable in the 1980s	Migration is seen as the inevitable outcome of the spread of capitalism. Migration is the only option for people once they are alienated from the land.
Gender studies	1990s	These emphasise that men and women differ in their responses to migration factors and that sex discrimination in the labour market has an important impact.

Table 4.1
A chronology of different ideas applied to the study of migration

The next case study illustrates two different types of migration, but both clearly belonging to the forced category.

TWO TYPES OF REFUGEE

Case study **15**

A refugee is defined by the UN as someone whose reasons for migration are genuinely to do with fear of persecution or death based on race, religion, nationality, political opinion or membership of a particular social group.

Examples of the circumstances generating recent refugee movements include:

■ **race** — whites fleeing the land-grabbing reforms of Mugabe's regime in Zimbabwe

- **religion** — the tension between Catholics and Muslims underlying the conflicts in Bosnia, Croatia and Kosovo
- **nationality** — the expulsion of Nepali-speaking people from Bhutan
- **political opinion** — citizens of Myanmar opposed to the military junta

When would-be refugees flee, they try to reach a nearby country as **asylum seekers**. An asylum seeker is someone who seeks to gain entry to another country by claiming to be a refugee. In most cases, they will be confined to a refugee camp. These camps are typically overcrowded and under-supplied. During their stay there, asylum seekers will be exposed to disease, crime and other harsh conditions. The individual stays in the camp until a country accepts their application for asylum and residency. The process of proving 'refugee' status is often very difficult. Countries that accept refugees will determine whether the applicant meets their own criteria for resettlement. The criteria might include age, health, education and occupational skills. This process can take a long time — as long as 8 or more years. Even after a refugee has been accepted as eligible for asylum and resettlement, there is often a very long wait for an available slot in the country that has accepted their application. If the application is refused, it is likely that the would-be refugee will be forcibly repatriated to their original country (Figure 4.8).

Figure 4.8
Immigrant pathways

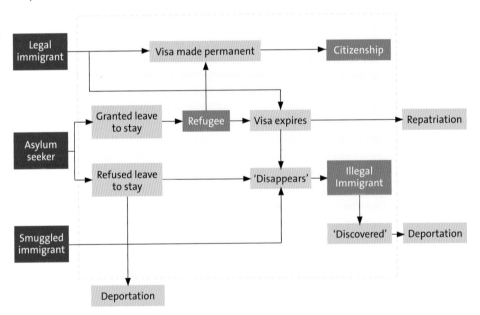

Annual figures released by the UNHCR showed that there were some 43.3 million refugees worldwide at the end of 2009 — that is, 0.6% of the global population. This was the highest number of people uprooted by conflict and persecution since the mid-1990s. At the same time, there were some 826,000 asylum seekers held in camps around the globe. The number of new asylum claims worldwide grew to nearly 1 million. South Africa received more than 222,000 in 2009, making it the single largest asylum destination in the world. Many of these asylum seekers came from neighbouring Zimbabwe and to a lesser extent from Malawi and Congo (Figure 4.9). The number of refugees voluntarily returning to their home countries had fallen to its lowest level in

20 years. Accurate statistics about the number of asylum seekers deported and forcibly repatriated seem not to exist. But there are many known current examples, such as the repatriation of Uzbeks from Kyrgyzstan, Chinese from Russia, Liberians from Ghana, and Afghans from a number of countries.

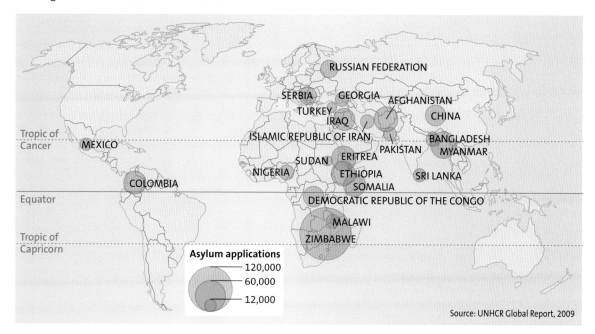

Source: UNHCR Global Report, 2009

Figure 4.9
Main countries of origin of new asylum seekers, 2009

Another type of refugee, not yet officially recognised by the United Nations High Commission on Refugees (UNHCR), is what researchers call the **environmental refugee**. These fall within the category known as **internally displaced persons (IDPs)**. They have been defined by others as 'persons who no longer gain a secure livelihood in their traditional homelands because of what are primarily factors of unusual scope'. Examples of such factors are desertification, natural hazards and, perhaps down the line, rising sea levels.

While it is true that the movements of environmental refugees are mainly confined within national borders, there are examples where environmental situations have been so severe as to trigger cross-border movements. These include the great famines. The total number of environmental refugees has been estimated as high as 25 million, a number that is less than the 'political' refugees currently of concern to UNHCR.

Refugees are a major component of the world's ever-growing migrant population. Genuine refugees need to be treated with respect and sympathy and ultimately offered asylum. However, the challenge for many governments today is to distinguish between bona fide asylum seekers and those opportunists who abuse the status as a way of circumventing immigration controls. See also Case study 35.

The next case study is an interesting one in terms of possible classification. Few would suggest that it is anything other than a voluntary migration, because the migrants did not have to leave. Admittedly, the decision to emigrate was conditioned by some strong push factors, and it so happened there were also some very persuasive opportunities at the destinations.

Emigration has emerged as one of the greatest challenges to South Africa since the end of apartheid in the early 1990s. Skilled South Africans are finding it increasingly attractive to work abroad, where their skills are eagerly snapped up. The outflow is seriously threatening the country's efforts to raise its rate of economic growth.

It is estimated that some 800,000 out of a total white population of 4 million emigrated from South Africa between 1995 and 2009. The chances of still more leaving are high. As in most developing countries, emigration from South Africa is essentially a brain drain, or exodus of skills. More than half of emigrants are professionals, semi-professionals, managers or executives. Typically they include doctors, lawyers, accountants and engineers, but their numbers are swelled nowadays by young people with IT skills, teachers, nurses, farmers and people with particular manual skills. It is estimated that more than 20% of South Africa's professionals have already left. Some 70% of skilled South Africans are currently considering emigrating. The causes of this exodus are many, with the main ones set out in Table 4.2. Some of these 'push' factors have mirror images that serve to reinforce the wish to migrate.

Table 4.2
South Africa: reasons for emigrating

Reason	Explanations
Violent crime	Reason cited by 60% of emigrants.
	South Africa now ranks alongside Russia and Columbia as one of the most dangerous countries in the world.
	During the 1990s, a total of 250,000 people were murdered; on average, 750,000 violent crimes were reported annually.
The economy	Reason cited by 10% of emigrants.
	Particular factors include:
	■ the huge devaluation of the South African rand
	■ the high rate of income tax
	■ high interest and inflation rates
	■ the increasing difficulty of many white people to find the kind of work that matches their skills
	The next three explanations are also discouraging investors.
Mbeki's two-nation approach	There was a shift away from Nelson Mandela's reconciliatory approach to whites to an emphasis on race under President Mbeki (1999–2008).
	Mbeki's belief was that South Africa consists of a rich white nation and a poor black nation and that economic sacrifices will have to be made by white people. The belief also appears to be held by Mbeki's successor, Jacob Zuma.
	Many white people are nervous about what those sacrifices might involve — for example, land grabbing and the displacement of whites by black landholders and farmers.
AIDS	South Africa now has one of the highest rates of HIV infection in the world.
	The strain this is putting on healthcare facilities, the debilitating impact of the disease on the workplace and the fear of contracting it are aspects that help to push the emigrants.
Zimbabwe	The crisis in neighbouring Zimbabwe has unnerved many white South Africans, who fear that the situation may spill over into their country. Many Zimbabweans have already fled to South Africa and their presence is having a destabilising effect.
The global village	In an era of globalisation, many emigrants simply wish to transfer their skills overseas to earn higher incomes.
	Some of these economic migrants eventually return, but for others the global village becomes their new home.

Just over 70% of emigrants move to five countries, all of them English-speaking developed countries with cultural similarities to, or historic ties with, South Africa (Table 4.3). The UK is the most popular destination because around 800,000 South Africans have, or can lay claim to, British passports. Britain allows visa-free visits for South Africans and offers 2-year working visas for South Africans who are under 27 years of age. There is no question that the influx of South Africans into the UK is helping fill gaps in the skills market: for example, within the National Health Service.

'South Africa is shedding skills at a worrying rate to its global competitors. The loss of each skilled professional results in the loss of as many as ten unskilled jobs. The destination countries have no compunction about creaming off skilled people from other countries,' says one South African official. So how is the South African government to react? Prohibit the emigration of mainly white workers? Train local black people to fill the gaps left by emigrating professionals? Encourage the immigration of foreigners? Another official has urged that 'If South Africa is to enjoy widespread long-term economic growth, it has to open its doors to foreign skills.'

Table 4.3 South Africa: destination of emigrants

Destination	% of emigrants
UK	25
Australia	18
USA	11
New Zealand	10
Canada	7
Western Europe (excluding UK)	25
Others (e.g. Switzerland, Israel, Namibia)	4

Points raised by this case study include:
- *The desire or need to emigrate is not confined to cheap, unskilled labour in search of work.*
- *A nation is only likely to lose its best workers when powerful push factors prevail.*
- *The movement of economic migrants can have serious cost implications for the source country, but considerable benefits for the receiving country.*
- *Should governments intervene to stop such costly movements?*
- *Is globalisation really a beneficial process when it allows labour to be drawn away from one country, only to be replaced by workers from another?*

It is ironic that whilst South Africa haemorrhages skilled labour, the country is the most popular destination for asylum-seekers from other African countries.

Two important points about the reasons for migrating need to be fully understood before moving on to consider the impacts of migration:
- The decision whether or not to migrate is largely conditioned by a number of factors that are personal to would-be migrants. These include their circumstances, their perceptions of opportunity and risk, and their values.
- In analysing the reasons behind the decision to migrate, it is easy to make assumptions that misinterpret motives. For example, the tabloid press in the UK prefers to see the current influx of immigrants from eastern Europe as being made up largely of people intent on 'sponging off the state'. In reality, it is much more likely that those people are driven to the UK by poverty, the need to find work (they have skills that are in demand) and the desire to start a new life. In short, migration is a topic open to bias and we need to be alert to this danger (see *Using case studies 9* in Part 5).

Impacts of migration

In Part 4, the focus was on the motives for migrating and how they could be subdivided into push and pull factors. The impacts of migration can similarly be subdivided into those at the **source area** and those at the destination or **host area** of the migration journey. There may also be a third type of impact, the effects of the move on the migrant. What sort of reception does the migrant receive? How easy is it for them to become assimilated into the host society?

Types of impact

The impacts of migration fall into a number of categories (Figure 5.1).

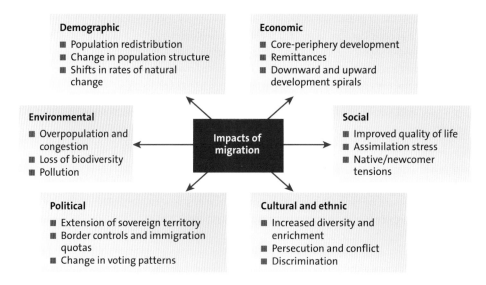

Figure 5.1
The impacts of migration

Demographic
- Population redistribution
- Change in population structure
- Shifts in rates of natural change

Economic
- Core-periphery development
- Remittances
- Downward and upward development spirals

Environmental
- Overpopulation and congestion
- Loss of biodiversity
- Pollution

Impacts of migration

Social
- Improved quality of life
- Assimilation stress
- Native/newcomer tensions

Political
- Extension of sovereign territory
- Border controls and immigration quotas
- Change in voting patterns

Cultural and ethnic
- Increased diversity and enrichment
- Persecution and conflict
- Discrimination

Demographic impacts

The most obvious outcome of migration is redistribution of population. Other consequences are related to the fact that migration is selective. For example, because it tends to involve people of reproductive age, migration has impacts on birth rates. For receiving or host areas, this means a double whammy — newcomers, plus a raised rate of natural increase. Conversely, for source areas, the loss of people is reinforced by a dip in birth rate and a decline in natural increase (see *Case study 31*).

Economic impacts

Much migration appears to be triggered by economic considerations, such as work opportunities, and wage and salary levels. What this means is that in-migration helps to boost locations that are already on the way up in terms of prosperity. That prosperity represents an irresistible attraction for many migrants, particularly those living in poorer and more deprived parts of the world. Out-migration can start, and aggravate, downward spirals of decline in source areas (see *Case studies 16 and 17*). Equally, the trickle back of **remittances** from emigrants can be a major source of income for family and relatives left behind.

Social impacts

The prospect of a better life is probably the single reason most mentioned by migrants when asked why they have moved. Happily, for most migrants the dream of a better quality of life is realised. However, there may also be social costs, such as the stress involved in adjusting to an unfamiliar and sometimes hostile environment. It is commonplace to find tensions between 'natives' and 'newcomers'. Often the latter form an underprivileged class obliged to do menial jobs and to occupy poor housing (see *Case study 20*).

Cultural and ethnic impacts

Throughout history, migration has played a key role in the spread of different cultures (particularly language and religion) around the world. Few would doubt that this spread has been enriching. However, conflict often arises when people of one particular cultural or ethnic background move to a location where they form an 'alien' minority. This is particularly so where intolerance prevails and where the immigrant and the host community feel in some way threatened (see *Case study 19*). The need for mutual security explains why immigrants of the same ethnic or cultural background are often found living in the same area.

Environmental impacts

Migration can give rise to adverse environmental impacts at both ends of the migration pathway. At the source end, it is the processes of abandonment, decline and neglect that scar the environment, whereas at the host end, the key processes are more likely to be congestion and the overstretching of resources (see *Case studies 22 and 34*).

Political impacts

The political impacts of migration can range from the growth of empires (like those of European nations during the seventeenth, eighteenth and nineteenth centuries) to the global spread of terrorist organisations such as al-Qaeda. The movement of people within a country often leads to governments having to face the complex issue of helping both **core** and **periphery** regions. Such internal movements can also alter voting patterns. A final point here is not so much about impacts as about the role that governments can, and do, play in both encouraging and discouraging migration (see *Case studies 13, 19, 20, 30 and 31*). Such political interference is made both possible and necessary by the fact that much migration involves crossing national frontiers. Tied up with all this are the complex issues raised by asylum seekers and refugees (see *Case study 35*).

It is important to note that, under each category of impacts of migration, there are normally both 'pluses' (benefits) and 'minuses' (costs), and that these costs and benefits often co-exist at both the source and the destination.

Impacts at source

There cannot be many examples of migration directly benefiting the source area other than in those instances of serious overpopulation, where out-migration provides a sort of pressure release. Otherwise the impacts are to be expected as largely negative. After all, any loss of people is likely to be accompanied by a loss of skills and enterprise.

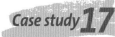

JAMAICAN MIGRANTS

There were large flows of migrants into Europe shortly after the end of the Second World War in 1945. Europe needed a lot of labour to repair the huge amount of bomb damage and to help the economy recover. But Europe was short of labour, because so many people, particularly men, had been killed. The situation was solved by encouraging migrant workers and their families to come to Europe, mainly from Africa, Asia and the Caribbean. The UK's postwar immigrants came mainly from colonies in the Caribbean, and from what had been the Indian Empire (India, Pakistan and Bangladesh). Immigration was encouraged by an Act of Parliament that gave all Commonwealth (ex-colonial) citizens free entry into the UK. The first ship to bring in immigrants from Jamaica docked at Tilbury (Essex) in June 1948 (Figure 5.2).

***Figure 5.2**
Immigrants from
Jamaica arriving at
Tilbury docks in June
1948*

It is estimated that during the 1950s and 1960s over a quarter of a million immigrants came from what had been the Indian Empire. Roughly the same number came from the Caribbean. Jamaica was not just the first country to supply migrants to postwar Britain — it probably became the biggest source country. Since the 1970s, when it became more difficult for Jamaicans to settle in the UK, large numbers have been emigrating instead to the USA and Canada. If nothing else, Jamaica has been a migration source region for over 50 years. It continues to lose some 20,000 migrants each year.

Impacts on Jamaica
One of the positive economic spin-offs of this continuing emigration is the increase in **remittances** — the money sent to family

TopFoto

TopFoto

Figure 5.3
Jamaica's scenic coastal landscape — what the emigrants are leaving behind

members back home in Jamaica. In 2005 remittances amounted to $1.65 billion (16% of Jamaica's gross national income). Between 1990 and 2005, this money helped to cut poverty in Jamaica by half.

An obvious negative economic impact of the emigration is that Jamaica is losing some of its best labour. There is a 'brain drain'. The island's economic development is being held back by this loss of skilled and more enterprising labour. Despite the loss, there is still high unemployment — a sure symptom of the slow rate of economic growth. High unemployment results in two things. It persuades even more people to emigrate. And it encourages people to turn to crime — in particular, there is gang violence related to the drugs trade. Clearly, this is a serious negative social impact.

Figure 5.4
Population pyramid for Jamaica, 2005

The majority of emigrants are young adults, with women outnumbering men. The loss of the more go-ahead women has a negative social impact. It is depriving Jamaican men of wives and depriving the country of both children and good mothers. The population pyramid (Figure 5.4) shows a number of features:

- a reducing birth rate — fewer babies
- a relative reduction of the population between the ages of 30 and 60
- fairly good numbers of elderly people, made up of those who have remained in Jamaica all their lives plus those migrants who have retired and decided to return home

It is not easy to identify the positive social impacts of the emigration. Can you think of any?

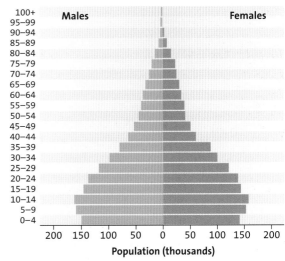

Source: US Census Bureau, IDB

This case study should be read in conjunction with Case study 20, *which looks at the impact of this emigration on the UK as a whole and on the migrants and their descendants.*

Impacts at destination

The next case study, involving another Caribbean island, is partly about motives, but more importantly it illustrates a number of the cumulative impacts on the host country.

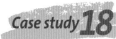

Case study 18 TRINIDAD: A MULTI-ETHNIC SOCIETY IN THE MAKING

Trinidad, a West Indian island just off the coast of Venezuela, has a population of 1.2 million made up of at least five officially recognised ethnic groups — African, Indian, Chinese, white and 'mixed'. This multi-ethnicity is the outcome of a long and varied migration history (see Figure 5.5).

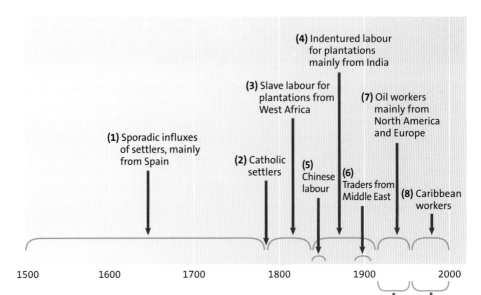

Figure 5.5
Trinidad: an immigration timeline

The numbered paragraphs in the following text correspond to the annotations in Figure 5.5.

(1) When Columbus discovered Trinidad in 1498, he found several tribes of Amerindian people inhabiting the island. Their total population was estimated at 35,000. The Spanish settlers, who followed Columbus, brought with them diseases (especially smallpox) and were harsh in their treatment of the Amerindians who, consequently, largely died out. However, even to this day, a small number of people in Trinidad claim to be descendants of the native people.

(2) The real development of Trinidad did not begin until the end of the eighteenth century when the king of Spain offered free land to citizens of any country friendly to Spain, provided they were Roman Catholic. This meant that most of the new settlers came from France, since England, Spain's other ally at the time, was mainly Protestant.

(3) The 'pull' of free land was strengthened by the opportunity it gave to the settlers to establish plantations growing a range of crops for the European market — mainly sugar, but also coffee, cocoa, cotton and citrus fruits. However, the plantations needed cheap labour. Before long, people were being rounded up in western Africa and forcibly transported (i.e. 'pushed') to Trinidad as slaves. Between 1783 and 1789, at least 10,000 arrived on the island. This immigration and the spread of plantations had a major impact on the environment as forests were cleared for cultivation and settlements. In 1797, Britain acquired the island as one of its colonies, and it remained so until it gained independence in 1962. Under British rule, the plantations expanded and all seemed well until 1834 when the British Parliament passed an act abolishing slavery. This caused much consternation among the plantation owners, but they eventually agreed to free all the slaves on the island. Some slaves left, but most stayed.

(4) As a consequence, the plantation owners were forced to find an alternative source of cheap labour. Eventually they turned to India. Under the indenture scheme, labourers were recruited for 5 or 7 years. At the end of this time, they could choose to be either repatriated or given some land and allowed to stay. By the time the Indian government banned emigration to Trinidad in 1917, the number of indentured workers had risen to more than 145,000. The significant difference was that immigration was now voluntary rather than forced. Many preferred to stay on. 'East Indians', as they are called today, now constitute around 40% of the total population, about the same as those of African extraction.

(5) Another group that found its way to Trinidad was the Chinese. They were brought in when there was a temporary halt in Indian immigration. The Chinese were not a great success as plantation workers. Soon they moved off to open small businesses of their own, forming a close-knit little community of active entrepreneurs. Today, their descendants account for around 1% of the population.

(6) Also involved in the commercial life of Trinidad today are people of Lebanese and Syrian origin. Their ancestors arrived mainly during the nineteenth and early twentieth centuries.

(7) The next phase in the migration story coincides with the first exploitation of oil (initially from the famous Pitch Lake) in the early part of the twentieth century. By the mid-1920s, oil was the number-one export. The expansion of the oil industry drew in more labour: technicians, mainly from North America and Europe, and unskilled labour from local sources and other West Indian islands.

(8) Today, the level of immigration is low. Probably up to 75% come from other Caribbean islands, apparently 'pushed' by high rates of unemployment and 'pulled' by the chance of work in service industries.

This long and varied immigration saga clearly explains the rich ethnic mix that characterises Trinidad society today. To their credit, the groups generally live in harmony. In fact, some 15% of today's population is classified as being of 'mixed' origins. However, when it comes to looking at the impacts of immigration, some differences between the ethnic groups begin to emerge (Figure 5.6).

The migration story of Trinidad is not just one of immigration. During the twentieth century, there were two significant exoduses of mainly unskilled workers: before and then after the Second World War (1939–45). Trinidad itself became a migration source

Figure 5.6
Trinidad: some impacts of immigration

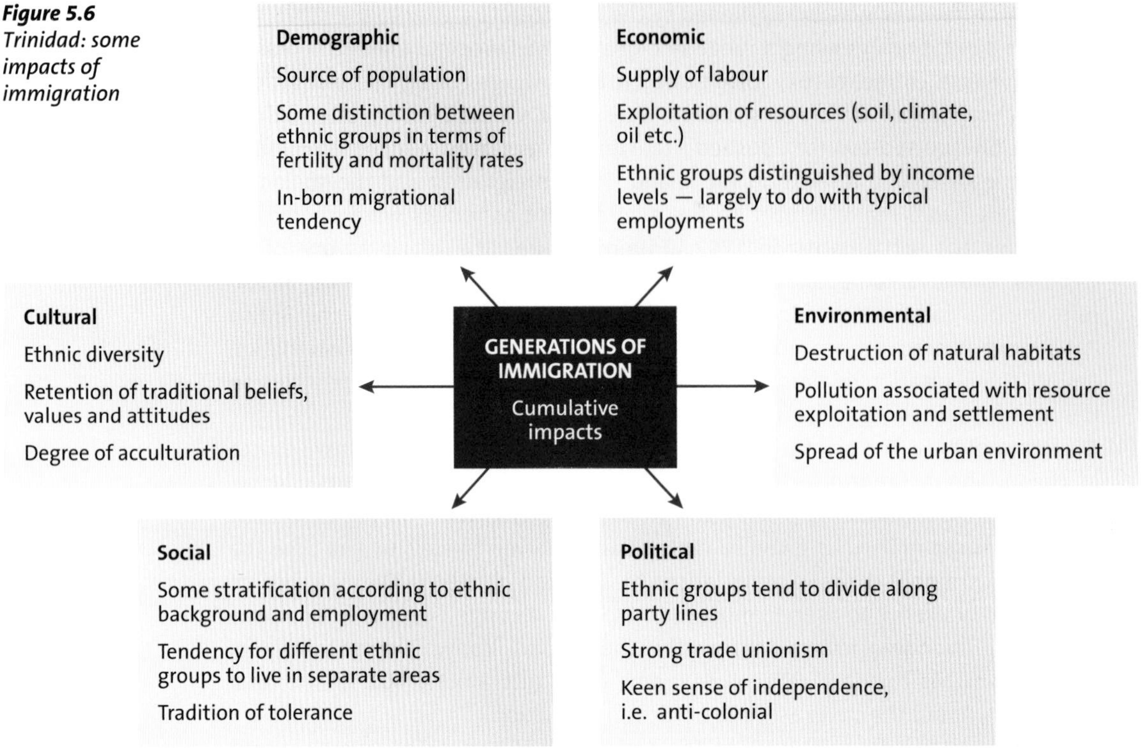

Demographic

Source of population

Some distinction between ethnic groups in terms of fertility and mortality rates

In-born migrational tendency

Economic

Supply of labour

Exploitation of resources (soil, climate, oil etc.)

Ethnic groups distinguished by income levels — largely to do with typical employments

Cultural

Ethnic diversity

Retention of traditional beliefs, values and attitudes

Degree of acculturation

GENERATIONS OF IMMIGRATION

Cumulative impacts

Environmental

Destruction of natural habitats

Pollution associated with resource exploitation and settlement

Spread of the urban environment

Social

Some stratification according to ethnic background and employment

Tendency for different ethnic groups to live in separate areas

Tradition of tolerance

Political

Ethnic groups tend to divide along party lines

Strong trade unionism

Keen sense of independence, i.e. anti-colonial

area. For example, during the period 1980–90, 52,000 people emigrated, 58% of them to the USA and 25% to Canada. In view of the historic links with the UK, it is surprising that less than 10% of these emigrants headed this way. Thirty years earlier, the situation was rather different (see *Case study 20*). However, as British economic interests have turned away from its former colonies, so the influence of the USA and Canada has grown.

No matter where the destination, most emigration has been 'pushed' by periodic slumps in the Trinidadian economy (particularly in the oil industry) and 'pulled' by employment opportunities and the perception that a 'better life' is to be found outside the island. Most emigrants are likely to be unskilled and obtain only low-paid jobs. The main concentration seems to be in hotels and in public services such as hospitals. Illegal immigrants often find that they are vulnerable to excessive exploitation by ruthless employers.

This case study shows that:
- *the modern population of Trinidad is almost totally migrant in its origins. Much of the explanation of how and why Trinidad is as it is today lies in a long and varied migration history.*
- *over the centuries, economic forces have been paramount in explaining the movement of people into and out of the country. In this respect, the Trinidad story is in no way exceptional.*
- *the ethnic diversity of Trinidad's immigrants is distinctive*
- *despite years of integration and living in reasonable harmony, the ethnic groups tend to remain distinct. Each tends to play a particular role in modern society, retaining much of its traditional culture. This is the most profound and enriching impact that immigration has had on Trinidad.*

Contemporary Case Studies

ECONOMIC MIGRANTS FROM EASTERN EUROPE

On 1 May 2004, membership of the European Union (EU) increased from 15 to 25. With the exception of Greek Cyprus and Malta, the geographical spread of the enlarged community was in an eastwards direction (Figure 5.7). The eight new member countries in this direction — the Czech Republic, Estonia, Hungary, Latvia, Lithuania, Poland, Slovakia and Slovenia — were all formerly part of the Soviet communist bloc. They are conveniently referred to collectively as the 'Accession 8' or 'A8' states.

The hype

One of the basic rights of EU membership is that all citizens are allowed free movement between member states. In the months immediately before May 2004, there were alarmist headlines in the tabloid press. They claimed that, because of this right, the UK was about to be overrun by hoards of economic migrants and 'benefit tourists' from A8 countries

So what did happen in May 2004? Did the floodgates open? In the first year some 150,000 A8 citizens registered in the UK, and as of September 2008 there were some 265,000 workers. Since then, it is estimated that the number has slowly been declining due to the impact of the credit crunch on the UK economy. Migrants have been returning in increasing numbers. It is estimated that over half of the early A8 workers came from Poland, followed by Lithuania, Latvia and Slovakia.

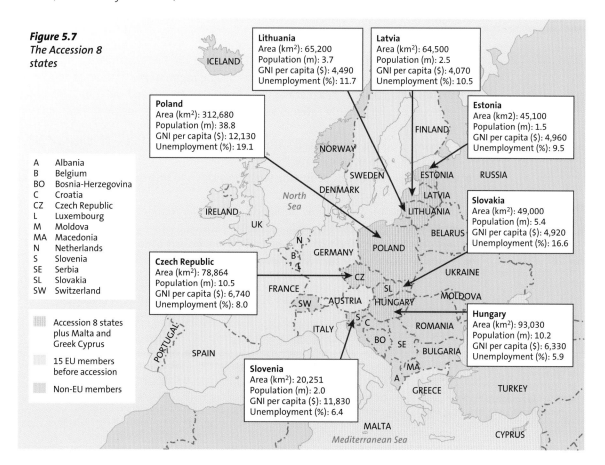

Figure 5.7
The Accession 8 states

Lithuania
Area (km²): 65,200
Population (m): 3.7
GNI per capita ($): 4,490
Unemployment (%): 11.7

Latvia
Area (km²): 64,500
Population (m): 2.5
GNI per capita ($): 4,070
Unemployment (%): 10.5

Poland
Area (km²): 312,680
Population (m): 38.8
GNI per capita ($): 12,130
Unemployment (%): 19.1

Estonia
Area (km2): 45,100
Population (m): 1.5
GNI per capita ($): 4,960
Unemployment (%): 9.5

Slovakia
Area (km²): 49,000
Population (m): 5.4
GNI per capita ($): 4,920
Unemployment (%): 16.6

Czech Republic
Area (km²): 78,864
Population (m): 10.5
GNI per capita ($): 6,740
Unemployment (%): 8.0

Hungary
Area (km²): 93,030
Population (m): 10.2
GNI per capita ($): 6,330
Unemployment (%): 5.9

Slovenia
Area (km²): 20,251
Population (m): 2.0
GNI per capita ($): 11,830
Unemployment (%): 6.4

A   Albania
B   Belgium
BO   Bosnia-Herzegovina
C   Croatia
CZ   Czech Republic
L   Luxembourg
M   Moldova
MA   Macedonia
N   Netherlands
S   Slovenia
SE   Serbia
SL   Slovakia
SW   Switzerland

Accession 8 states plus Malta and Greek Cyprus

15 EU members before accession

Non-EU members

The myths

It is clear that there are people in the UK who have taken a distinctly negative attitude towards these east European workers. It is claimed that they are depriving UK workers of jobs and taking advantage of UK benefits. This last accusation is levelled particularly at migrants from Bulgaria and Romania. Four keys facts seem to be ignored:

■ These workers, particularly those from A8 countries, contribute to the UK's gross domestic product.

■ The jobs that many of these economic migrants take up are mainly low-paid. Such jobs are often avoided by UK workers. One of the ironic features of the UK today is the coincidence of low levels of unemployment and high levels of benefit claims from UK citizens. A chief executive of a labour recruitment company recently put it this way: 'Some of these people [on benefit] would rather work fewer hours and...so they retain their benefits. Foreign workers are happy to work between 30 and 50 hours a week.'

■ The east European economic migrants have a strong work ethic, which can directly benefit employers. Sadly, there are employers who have unfairly exploited this quality (see below).

■ Curbs have been in place since 2004 that restrict the access of A8 workers to benefits (including healthcare) and housing benefits.

Hopefully, enough has been written to dispel the two myths about migrants 'stealing' UK jobs and 'milking' the UK's benefit system. However, it has to be admitted that there are some serious issues associated with this economic migration from eastern Europe.

The abuse

There have been various exposés in the media about the bad treatment of immigrant workers, particularly at the hands of the so-called **gang-masters**. A report produced by the Trades Union Congress has catalogued a long list of abuses and hardships, including:

■ employers paying hourly rates lower than promised, or lower than those paid to British workers

■ excessive working hours, often with no rest day allowed, no breaks between shifts and no enhanced overtime rate

■ employers and agencies deducting what they claim to be income tax on wages that are below the tax threshold

■ personal documents, such as passports, unlawfully retained by gang-masters and employment agencies

■ very poor housing conditions – substandard and grossly overcrowded

■ excessive charging for such accommodation

Figure 5.8
The push, pull and related issues of A8 migrant workers in the UK

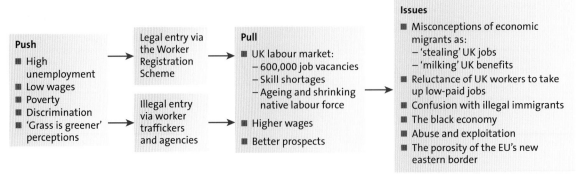

- the hostile tone of some press reporting
- overt xenophobic actions such as petrol bomb attacks on homes and hostels, verbal abuse and taunting in the streets

Having read this abbreviated list, you might be wondering why, in the first place, these workers come to the UK and why they stay for as long as they do. The short answer is that this economic migration is the outcome of a powerful 'push' situation aligned with a fairly attractive 'pull' scenario. Figure 5.8 seeks to detail this and to set out some of the related issues. What also needs to be stressed is that what we have here is probably a case of temporary migration. It is assumed that eventually most of these workers will return home. The fact is, however, that we simply do not know for certain. It is amazing to learn that the Home Office only keeps detailed information on immigration. People leave the country virtually unnoticed and unrecorded.

This case study illustrates the vital point that host country views of immigrants are coloured by personal perceptions, misconceptions and plain prejudice. Reference to Figure 5.1 on page 56 may help you to put this particular migration in context.

Impacts on migrants

The last two case studies in this part of the book put the spotlight on the immigrants themselves and the challenge of adjusting to their new homelands. What happened to those immigrants from Trinidad and other Caribbean countries, as well from Africa and Asia, who entered the UK after the Second World War?

THREE GENERATIONS OF UK IMMIGRANT .

Case study **20**

There was a substantial influx of Asian and African-Caribbean immigrants into the UK during the 1950s and 1960s. A significant proportion was made up of young adults who have since reared families and become grandparents. Data from the 2001 census have given the most detailed picture of ethnic Britain. The census form allowed people to describe themselves as mixed race, rather than having to choose white, Asian, black or Chinese ethnicity, while more questions were asked about people's homes, family set-up and overall health. People from ethnic minority groups now make up almost 8% (4.6 million) of the population, an increase of 53% over the 1991 census. Figure 5.9 shows the breakdown of the UK's ethnic population. The fact that 15% of that population declared themselves as being of 'mixed race' provides evidence of integration. So too does the fact that half of that ethnic minority population was born in the UK. It emphasises the crucial point that these people are part of the community and not foreigners.

Figure 5.9
The composition of the UK's ethnic minority population, 2001

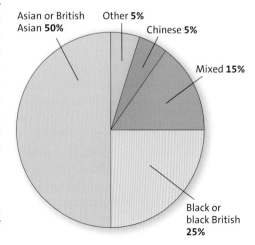

Asian or British Asian **50%**   Other **5%**
Chinese **5%**
Mixed **15%**
Black or black British **25%**

Inequality

It is a widely held belief that first-generation immigrants often suffer in their new homeland, particularly in terms of having to take low-paid jobs and occupy poor housing. In contrast, the offspring of immigrants (i.e. the next generation), since they are citizens of the country in which they were born, are seen as typically moving off the breadline and becoming better off. However, the results of the 2001 census challenge these beliefs. They show that black, Asian and other ethnic minorities are twice as likely to be unemployed and are half as likely to own their home. Their risk of poor health is double that of white Britons. The results also show that the proportion of Muslim children living in overcrowded housing is more than three times the national average; that they are twice as likely to live in a house with no central heating; and that children from Pakistani and Bangladeshi families suffer twice as much ill-health as their white counterparts. Residential segregation on the basis of ethnicity, particularly in inner suburbs, remains the rule. Is that by choice or force of circumstance?

In short, the census results indicate that the gap between the richest and poorest in the UK has never been wider. More worrying is the fact that the widening gap has an ethnic dimension. One researcher has described the situation as follows:

> Some people from ethnic minorities are moving up and becoming wealthier, but for some children the future is bleak. They are getting the worst start in life and the problem becomes a vicious circle because of the correlation between poor housing, poor education and future criminality.

Figure 5.10
'Do you feel British?'

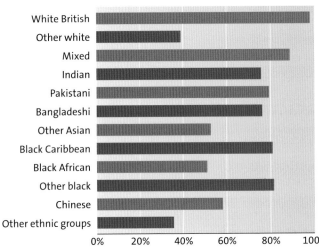

Proportion from each ethnic group to answer 'yes'
(British includes English, Scottish, Welsh or Irish)

It is difficult to know exactly what needs to be done to ensure a fairer future for all. Successive governments have set up a range of programmes to tackle discrimination, poverty and inequalities, but they simply do not seem to be working. Instead, it looks as if generations of the UK's ethnic minorities will continue to grow up in deprivation.

However, all is not gloom; there is at least one ray of hope. A recently published survey has shown that a clear majority of people from all the main ethnic minorities identify themselves as British citizens (Figure 5.10). Feelings of 'Britishness' are strongest among people of mixed race.

The future

The point needs to be made that the UK today is not just home to three generations of New Commonwealth immigrants. As Figure 5.11 shows, the UK also continues to attract significant flows of immigrants from other parts of the world. Equally important, it illustrates the two-way nature of migration, with the UK also acting as a source.

Looking to the future, a report in 2010 has predicted that in 40 years' time, the proportion of black, Asian and other ethnic minorities will rise from 8%, as recorded in the 2001 census, to 20% by 2050. It is also predicted that by the latter date the total population of the UK will reach nearly 78 million — it was 59 million in 2001. The point to be stressed is that the rise in proportion of ethnic minorities will be due more to

Contemporary Case Studies

COMINGS ➤

EU workers
- Freedom of movement within the EU encourages large flows, particularly from 'southern' member states (Portugal, Spain, Greece)
- Immigrants from the EU account for only 13% of UK immigration

Non-EU workers with permits
- Mainly from eastern Europe, particularly countries joining the EU in 2004
- 'Work-related' moves account for 26% of UK immigration
- 34% of all UK immigrants come from outside the EU and Commonwealth

Illegal immigrants
- Mainly employed in the black economy, including Chinese cocklers, Albanian prostitutes, east European farm gangs, etc.
- No official statistics; estimates put the flow in the order of tens of thousands

Asylum seekers
- Today mainly from the Middle East (Iraq, Iran, Afghanistan) and China
- Significant flows from Africa (Somalia, Zimbabwe, Sierra Leone)
- Influx from former Yugoslavia now easing
- Account for 15% of all UK immigrants

Family and friends
- Relatives of families already resident in the UK
- Includes the 'victims' of 'forced' marriages
- 'Accompanying/joining' accounts for 18% of UK immigration
- New Commonwealth citizens are the largest single immigrant group (32%)

Students
- Mainly from Europe and Commonwealth countries for degree and English language courses
- 'Formal study' accounts for 18% of UK immigration

Returning ex-patriates
- Often due to death of partner, disillusionment, end of contract
- Significant numbers from within the EU (France and Spain)
- 22% of UK immigrants hold British passports

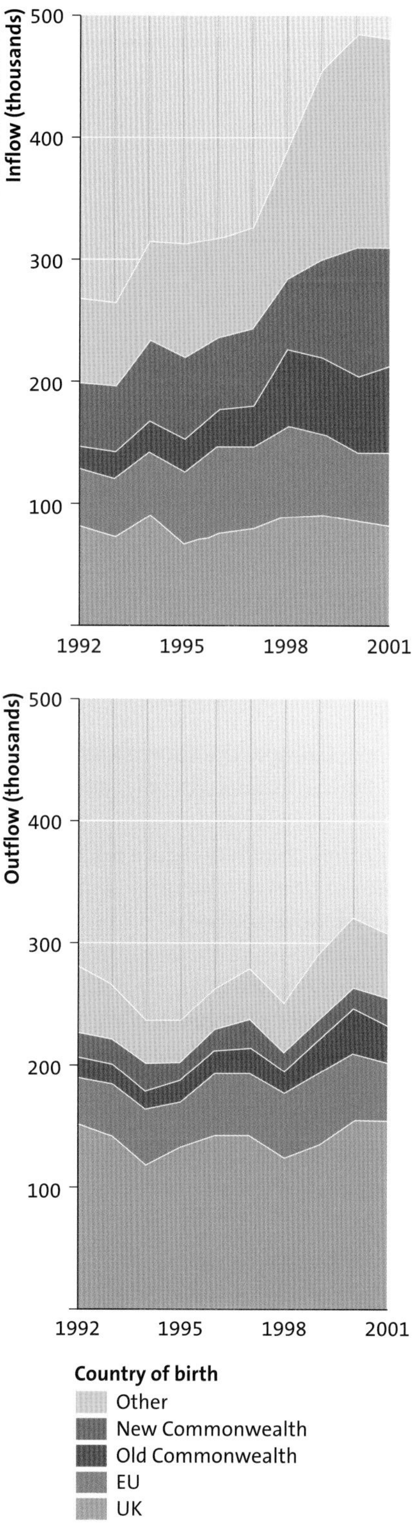

GOINGS ➤

Overseas postings
- Mainly professional people and technicians taking up long-term contract work, often in LEDCs
- This category also includes the 'brain drain' to other MEDCs
- 30% of emigrant moves from the UK are 'work related'

Discontents
- People dissatisfied with their quality of life, the cost of living, job prospects, etc.
- France and Spain may be popular destinations
- There are no statistics; they are included in the 'other reasons' category accounting for 34% of all emigration

Refused asylum seekers
- Most of these are economic migrants who are 'removed' and sent back to homes in eastern and southeastern Europe, the Middle East and New Commonwealth countries
- Account for less than 5% of emigrants

Returnees
- Normally first-generation immigrants who wish to retire to their roots, having made sufficient money working in the UK
- They account for a significant proportion of the 16% of emigrants bound for Commonwealth countries

Students
- Includes 'gap year' students and undergraduates going to universities in North America and Old Commonwealth countries
- They account for around 5% of UK emigrants

Country of birth
- Other
- New Commonwealth
- Old Commonwealth
- EU
- UK

Figure 5.11
UK international migration, 1992–2001

natural increase in a relatively young adult population than to immigration, especially if immigration is controlled.

In terms of future integration, one of the authors of the report has said: 'Our results suggest that overall we can look forward to being not only a more diverse nation, but one that is far more spatially integrated than at present.' More specifically, he suggests that: 'At a regional level, ethnic minorities will shift out of deprived inner city areas to more affluent areas, which echoes the way white groups have migrated in the past.' However, as if in contradiction, he adds that: 'In particular, black and Asian populations in the least deprived local authorities will increase significantly.' That hardly suggests an integrated society! Neither does the fact that ethnic minorities have scarcely made an appearance in the UK's rural areas.

This case study shows that, despite well-intentioned legislation, ethnic minorities in the UK still do not enjoy the same range of opportunities as the white majority. There are two related discussion issues:
- *What else might be done to ensure equality?*
- *Is the loss of ethnic identity the price that has to be paid for achieving equality and integration?*

---

9 *Using case studies*

Perception is an important aspect of migration. It arises in two different, but related, contexts. First, there is the perception by the migrants themselves, particularly of the push and pull factors impinging on them. In many cases, it is the perception and valuation of opportunities at the potential destination that are particularly important. Second, there is the perception of migrants by people living in the source and the host areas. What do they think about those who are leaving? Perhaps more critically, what do they think of the incomers? Here is a perception exercise for you to work on.

(a) For each of the incoming groups shown in Figure 5.11:
   (i) put yourself in the place of one member and identify your likely perception of the upside and downside of a move to the UK
   (ii) try to identify the likely attitudes of those remaining behind towards these emigrants
(b) For each of the outgoing groups shown in Figure 5.11:
   (i) put yourself in the place of one member and identify your likely perception of the upside and downside of living in the UK
   (ii) try to identify the likely attitudes of people in the host country towards this group of migrants
(c) Compare your responses to (a) and (b). Are there significant differences or similarities?

---

10 *Using case studies*

Question

With reference to specific examples, assess some of the benefits and costs of international migration to both source and host countries under the subheadings:
(a) environment
(b) economy
(c) society and culture

The citizens of most developed countries are living longer. In the UK, average life expectancy for women is now 81 years, and for men it is 76 years. Most people can expect to enjoy 10 or more years of living on a pension beyond the present official retirement ages of 65 (for men) and 60 (for women). With this prospect, more and more people are moving after they have retired. They are doing this for a number of reasons:
- it is no longer necessary to live close to what was their place of work
- to downsize into a smaller home
- to sell their home for something cheaper and use the difference in price as a pension
- to move into a quieter, calmer and more attractive environment

Three main types of retirement migration may be recognised:
- local — where people stay in the same locality, but move house
- regional — where people stay within the UK but move to what they think is a more attractive location
- international — where people make the bold decision to move to another country

RETIRING TO SPAIN: DREAM OR NIGHTMARE?

Case study 21

Surveys of retirement migration often reveal that weather is a factor in deciding on a retirement destination. Old people feel the cold, and heating a house in the UK is expensive. It is for this reason that increasing numbers of Britons are retiring abroad. Most head in the direction of the Mediterranean, but some go even further south to places such as the Caribbean. Adding to the pull of such locations is the relatively lower cost of housing. An attraction of staying within the EU is that basic healthcare is free to UK citizens — and the destination country's health service might be even better than the NHS in the UK.

Peter Titmuss/Alamy

Figure 5.12
People who retired to Spain probably gave a lot of thought to the view and how far they would be from the beach, but not to falling ill or the problems that age can bring

With retirement looming, you sell up in Britain, buy your dream villa in Spain and set off to live out your golden years in the sunshine. The idea is so tempting that three-quarters of a million Britons have already done just that. In some areas of the Spanish coast, one in ten residents is now of British origin.

Many of these retirement migrants no doubt gave a lot of thought to the view from the villa and how far they would be from the nearest golf course or beach (Figure 5.12). Many probably did not think about falling ill or the increasing problems that age can bring. The consequences can be catastrophic. Some migrants are forced to sell up and come back to the UK, often with no savings or property to fall back on. That is partly because the huge influx of elderly settlers is putting a massive strain on Spain's health service. Some Spanish doctors are refusing to treat anyone who does not speak Spanish unless they have an interpreter present. They fear they might make a wrong diagnosis and be sued.

Another problem is that the Spanish system of caring for the elderly is different from that in the UK. In Spain, it is traditionally a family responsibility. As a result, there are very few services for the elderly. Most British couples retiring to Spain have no family living close by to help them, should they fall ill. In the UK there is after-care treatment when the elderly come out of hospital. There are old people's homes, meals-on-wheels and care in the community. A British Consul in Spain recently said:

> Sometimes a partner has died and the other is too old or too ill even to go out and buy food. Some have not budgeted their pensions properly and are living in extreme poverty. British retirees need to realise that some European countries do not have the welfare provision available in the UK.

Contemporary Case Studies

Retiring to another country is very different from holidaying there. The key to a successful retirement move is knowing exactly what to expect. For some, the reality will not live up to the dream.

This case study on retiring to Spain illustrates the following points:
- *There are risks involved in moving to a foreign country, particularly for retirement.*
- *Migrants often tend to be blinkered — they see only the attractions of their destination and ignore the negatives. Perceptions can often deceive.*

11

Using case studies

Question

Much of what we read, whether it be a textbook or magazine article, has an in-built slight bias that reflects the values and perceptions of the writer. However, there are contentious aspects of population and migration in which there is scope for bias of a more sinister kind. This seeks to persuade the reader to see things in a particular way, which may be removed from reality.

Read the following extract taken from the editorial column of a tabloid newspaper shortly after the enlargement of the EU in 2004:

> The flood of Britons moving to live in Spain is the latest sign of growing disillusionment with British life in the twenty-first century. The UK may be a 'land of milk and honey' for many illegal immigrants and spoof asylum seekers, but not so for the increasing number of Britons who think that the 'good life' is to be found elsewhere.
>
> Our report into the life enjoyed by expatriates from the UK shows that Spain offers more than just sun and sea. It is not just a destination for pensioners and second-homers. It also offers higher salaries, safer streets, better schools and state-of-the-art hospitals. Even some of the immigrants coming here are only using British citizenship as a stepping-stone to somewhere better.
>
> Bringing up the standard of living in all new EU countries is something to be encouraged. It will reduce the incentives for migrants from eastern Europe to head to Britain. Equally, the UK government cannot just sit back and allow the exodus of some of our most skilled people to continue unchecked. Migration both into and out of the UK needs to be checked.

Read *Case studies 13, 19* and *21*, research the factual basis of this extract and expose its bias.

Guidance

It may be helpful to use some of the information in Figure 5.11 on page 67.

Population and resources

Consumption

People are first and foremost consumers. The very survival of the human race requires basics such as air, water, space, food and shelter. Such survival needs are, in effect, resources:

- natural (e.g. water, soil)
- human (e.g. enterprise, education)
- renewable (e.g. solar energy, forests)
- non-renewable (e.g. fossil fuels, minerals)

The rate at which resources are consumed (consumption) is strongly influenced by two processes: population growth and development. We can think of population growth, resources and development as being involved in a pattern of triangular relationships that are reciprocal or two-way (Figure 6.1).

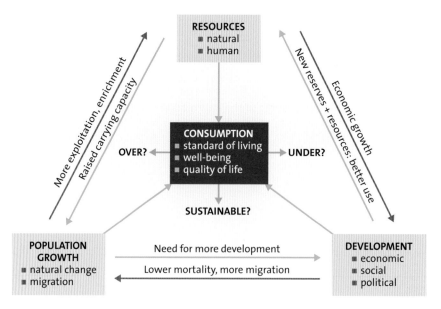

Figure 6.1
The consumption triangle

We can call this set of relationships the 'consumption triangle'. Figure 6.1 shows that while population growth enriches human resources (such as labour and enterprise), it also raises the demand for, and the exploitation of, resources. A knock-on effect is to encourage economic growth and therefore development. Development may be expected to impact on population growth by reducing mortality (better health-care) and encouraging in-migration (economic asylum seekers).

A 'circuit' of relationships can also be detected going in the opposite direction. For example, population growth creates a demand for devel-opment. More development increases resource exploitation. Equally, new technology (a part of development) could lead to more efficient use of resources, as well as to the discovery of new reserves and resources. A resource base enlarged in these ways will raise the carrying capacity of an area and therefore its ability to support more population growth.

Figure 6.2 More people means more consumption: Bullring shopping centre, Birmingham

Sitting in the middle of the triangle is consumption. While the rate of population growth is a critical factor here (more people means more consumption), so too is the ability of development and the resource base to sustain rising levels of consumption. If that capability exists, there is the promise of rising levels of human welfare, as reflected in a range of measures such as standard of living, wellbeing and quality of life. If that capability does not exist, deterioration in the condition of a population seems inevitable. Therefore, it is the nature of the balance between resources, development and population growth that is of paramount importance. Broadly speaking, three different states of 'balance' may be recognised, each at two levels — national and individual:

- **Over-consumption.** At a national level, over-consumption occurs when, and where, population growth races ahead of the current availability of resources and the present level of economic development. It involves low standards of living, and its symptoms include poverty, malnutrition, disease and environmental damage. However, at a personal level, over-consumption involves a totally different scenario. Among its more obvious symptoms are affluence, a profligate use of resources and obesity.
- **Under-consumption.** At a national level, under-consumption is relatively rare today. It exists where resources and development could support a larger population without any serious depletion of known reserves. In stark contrast, under-consumption at an individual level is widespread. Its symptoms include those characteristics mentioned above in connection with national over-consumption: namely, poverty, malnutrition, disease and short life expectancy.

■ **Sustainable consumption.** This is a much sought-after state of balance, but one that could prove difficult to achieve nationally. It requires a stable population, innovative development and the consumption of renewable rather than non-renewable resources. It is a situation in which the living standards of future generations are not threatened by present consumption. However, as individuals, each of us can do a great deal to make our lifestyle more sustainable by reducing our **ecological footprint**. Actions we can take include, among others, recycling, using public transport rather than the car and eating organic food.

Case study 22 — THE ECOLOGICAL FOOTPRINT

The ecological footprint (EF) is a measure of a population's consumption of renewable natural resources. It is usually calculated in terms of the total area of productive land and sea required per person to meet their food, energy, raw material, water and waste disposal needs. The 'depth' of the footprint is conditioned by three key factors:
■ the rate of population growth
■ the levels of development and consumption
■ the nature of available technology

Figure 6.3
The global ecological footprint

The EF of a population can be calculated at a range of spatial scales from global and regional to national and local.

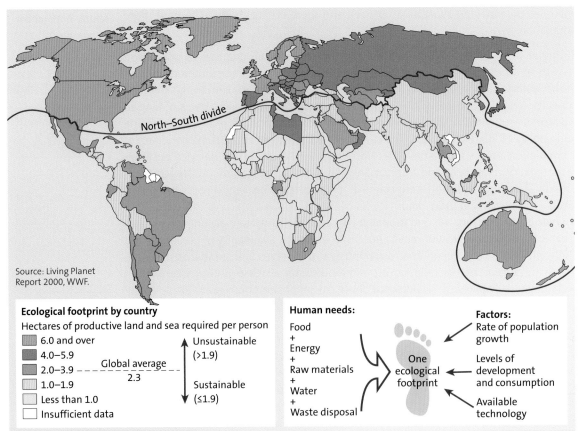

Source: Living Planet Report 2000, WWF.

Ecological footprint by country
Hectares of productive land and sea required per person

- 6.0 and over — Unsustainable (>1.9)
- 4.0–5.9
- 2.0–3.9 — Global average 2.3
- 1.0–1.9 — Sustainable (≤1.9)
- Less than 1.0
- Insufficient data

Human needs:
Food
+
Energy
+
Raw materials
+
Water
+
Waste disposal

One ecological footprint

Factors:
Rate of population growth

Levels of development and consumption

Available technology

The Earth has about 11.4 billion hectares of productive land and sea space — about a quarter of its surface area. Shared by a global population of 6 billion, this equates to 1.9 ha per person. According to a recent report by the World Wide Fund for Nature (WWF), while the EF of the average African and Asian consumer was less than 1.4 ha per person in 2000, the average western European's footprint was 5.0 ha and the average North American's 9.6 ha. The mean EF of the global consumer was 2.3 ha per person. This is 20% above the Earth's biological capacity of 1.9 ha. This means that the human race now exceeds the world's ability to sustain its consumption of renewable resources. Currently, the human race is only able to maintain this 'global overdraft' on a temporary basis by eating into the Earth's capital stocks of forest, fish and fertile soils. We are also dumping our excess carbon dioxide into the atmosphere. Neither of these two activities is sustainable in the long term. The only sustainable solution is to live within the biological capacity of the Earth. If we do not, the outlook can only be one of catastrophe.

The current trends show that the human race is moving away from achieving the minimum requirement for sustainability, not towards it. The global EF grew from about 70% of the Earth's biological capacity in 1960 to about 120% of its capacity in 2000. Projections suggest that the EF is likely to grow to around 200% of the Earth's biological capacity by the year 2050. However, it is unlikely that the Earth would be able to run an ecological overdraft for another 50 years without some serious backlashes in terms of human welfare, economic development and the environment.

Can we stop the footprint getting deeper?

If the human race is to follow a sustainable development pathway, it will involve making changes in four fundamental ways. We must:

- improve the efficiency with which resources are used in the production of goods and services
- reduce the 'widening gap' between high- and low-income countries, particularly by reducing the high consumption levels of the former
- protect, manage and restore natural ecosystems and so maintain (or even enhance) biological diversity and biological productivity
- control population growth through the promotion of contraception, education and healthcare

No nation can go it alone. Achieving a sustainable future requires everyone to play their part in helping to reduce the global ecological footprint. What are you prepared to do about it?

12 **Question**

Which of the four changes bulleted in *Case study 22* is most crucial in finding a sustainable development pathway?

Guidance

This question is for you to debate with other members of your group. Organise yourselves so that one or two students make a presentation in support of one of the four changes. If you wish to convince others, you will need to marshal supporting examples. Perhaps your teacher might act as judge and jury!

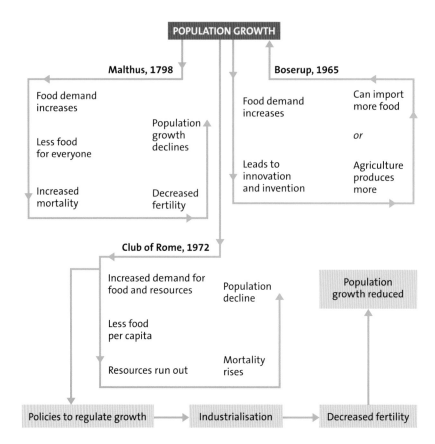

Figure 6.4
Three theories about the relationship between population growth and resources

Over the years, three main theories have been put forward about the relationship between population and resources (Figure 6.4). They can be summarised as follows:

- Malthus (1798) argued that food supply acts as a ceiling to population growth and that population growth takes place at a faster rate than the increase in food supply.
- Boserup (1965) argued that population growth stimulates increased food supply, either by improving the productivity of farming or by importing supplies.
- The Club of Rome (1972) predicted that if the current trends in population growth, industrialisation, food production, pollution and resource depletion were maintained, the limits to global population growth would be reached within 100 years. Population would then decline. Before that point, policies could be introduced to decrease fertility and therefore lower the rate of population increase.

Over- and underpopulation

The two issues that recur under the heading 'consumption' are **overpopulation** and **underpopulation**. It might be said that overpopulation (in the sense of a mismatch between the number of mouths and available food) is more of a global problem. For this reason, perhaps we should look to the United Nations and its agencies to set the agenda and policies. On the other hand, underpopulation is a problem for a relatively small number of countries. This is said to exist where resources and development could support a larger population without any lowering of living standards (e.g. Canada and New Zealand), or where a population is too small to develop its resources effectively (e.g. pioneer and wilderness regions). However, recent events are beginning to define a new underpopulation scenario.

13 *Using case studies*

Question

With the use of examples, explain why the relationship between population and resources is so important.

Guidance

First, explain briefly the threefold relationship: underpopulation, **optimum population** and overpopulation.

The root of a satisfactory answer lies in identifying the consequences of the unsustainable situation when population exceeds the carrying capacity. Consequences in the natural environment include:

- soil erosion
- deforestation
- air and water pollution

Consequences in the human environment include:

- reduced per capita food supply, leading to malnutrition
- substandard housing
- underemployment
- overstretched services
- political and social unrest
- impaired quality of life

Many of these symptoms are supported by the example of Ethiopia (*Case study 9*).

The discussion might move on to make the point that most of these negative outcomes gradually disappear as population is slowly brought into balance with resource availability and consumption. Use might be made of Chile (*Case study 23*) as an example of a country that has the opportunity of doing this by simply 'tweaking' its distribution of population.

CHILE: UNDER- OR OVERPOPULATED?

Case study **23**

Chile is one of the most weirdly shaped countries in the world. It occupies a thin strip of land, rarely more than 200 km wide, which stretches 4,640 km down the west coast of South America from north of the Tropic of Capricorn to Cape Horn (55°S). As a consequence, the climate ranges from hot desert in the north to cool temperate in the south. Southern Chile is one of the world's wettest and stormiest regions. Only 2% of Chile's land area is suitable for growing crops, 17% for livestock and 11% for forestry. The remaining 70% is covered by deserts or mountains (the Andes dominate) and is considered unproductive for agriculture, but suitable for mining.

Persons per km²

- ▮ > 79.9
- ▮ 40–79.9
- ▮ 20–39.9
- ▮ 10–19.9
- ▮ < 10

PERU

BOLIVIA

TARAPACA

ANTOFAGASTA

CHILE

ATACAMA

Pacific Ocean

COQUIMBO

ARGENTINA

VALPARAISO

SANTIAGO

O'HIGGINS

DEL MAULE

BIO-BIO

DE LA ARAUCANIA

DE LOS LAGOS

0   km   500

N

AISEN

MAGALANES ANTARTICA

Atlantic Ocean

Figure 6.5
Chile: the distribution of population, by region, 2002

With a population of just over 15 million living at a mean density of 20 persons per km², Chile ranks as one of the most sparsely populated countries in the world. Does this necessarily mean, however, that it is in the enviable position of being a country of underpopulation and under-consumption?

Distribution of population

If we look at the distribution of population in Chile, we might readily conclude that the central regions are overpopulated (Figure 6.5). In the two contiguous regions of Santiago and Valparaiso, just over half of Chile's population is living on a mere 4% of the land area, at a mean density of 240 persons per km² (six times the national average). Of the various climates that Chile has to offer, the Mediterranean type found here in the middle of the country is the most attractive. However, high levels of atmospheric and water pollution indicate that the concentration of so many people and so much of the country's economic activity in the central area is seriously degrading the environment. Air pollutants become trapped within the Central Valley cut deep into the Andes, making Santiago one of the world's most polluted cities.

If we look outside the central area and focus instead on the remainder of the country, with its mean density of 10 persons per km², we might conclude that Chile is, after all, an underpopulated country.

Be careful here. The key to understanding overpopulation is not population density but the number of people in an area relative to its resources and the capacity of the environment to sustain human activities. A country becomes overpopulated when its population cannot be maintained without the rapid depletion of its non-renewable resources and without degrading the capacity of the environment to support the population. In short, if the current human occupants of a country are clearly degrading its long-term carrying capacity, that country is overpopulated.

Chile and the global economy

Like most of the rest of the world, Chile is being drawn into the global economy. It now meets a significant proportion of the global demand for copper and nitrates. These are its chief exports (accounting for around half of all exports by value in 2001). Other exported minerals include heavy bitumen and iron ore. This dependence on selling non-renewable resources does not bode well for a sustainable future. There is also much environmental damage to take into account.

The situation with respect to renewable resources is not much better. Large exports of fish and crustaceans (fresh and prepared) are depleting the stocks of Chile's once-rich waters. Still worse is the large-scale deforestation caused by big exports of timber, wood pulp and paper. While the rising production and export of fruit and wine may look to be a promising development, it has required taking over land once used for subsistence purposes (both crops and meat). Remember that agricultural land is a scarce resource in Chile. So while Chile exports food and drink products worth $5 billion, its imports in the same category amount to $1 billion. The irony is that Chile is quite capable of producing many of those imported products — indeed, it used to do so. It seems perverse that Chile is not doing more to meet its increasing food needs by exploiting its own range of climates. In theory, this climatic range should allow it to produce almost anything, from tropical crops to cool temperate livestock.

The future?

Two indicators suggest that Chile may well be living dangerously. Its consumption pressure is currently put at 3.26 (three times the global average) and its ecological footprint at 3.2 (one and a half times the global average).

There is no doubt that since the political troubles of the 1970s and 1980s, the economy of Chile and the standard of living of its people have improved. It is now recognised as a 'middle-income' economy. However, both these achievements have involved much higher levels of consumption. There are serious warnings that Chile may be reaching the critical thresholds of over-consumption and overpopulation, particularly given the current population growth rate of 1.7% per annum.

This case study underlines the following points:
- *Population density is a poor indicator of overpopulation.*
- *Globalisation raises consumption and often this will involve a country exploiting its resources for the benefit of foreign rather than domestic consumers.*
- *Renewable resources, such as fish stocks and forests, are vulnerable to over-exploitation.*
- *Overpopulation exists where there is serious depletion of resources and environmental degradation.*
- *The speed at which a country reaches the threshold of overpopulation is influenced by the population growth rate.*
- *The speed at which a country reaches the threshold of over-consumption is influenced by the economic growth rate.*

THE WORLD'S MOST POPULATED COUNTRIES

Case study **24**

In August 1999, the world's population passed the 6 billion mark. Less than a year later, India joined China as the second country with a population over 1 billion. Projections indicate that in 50 years' time, India will surpass China as the world's most populous nation, when it is expected to have 1.5 billion inhabitants. The basic reason for the change in ranking is that while China has a family planning programme (see *Case study 29*), India's attempt to put one in place has been a colossal failure; in India today, a baby is born every 2 seconds. Its population is increasing by 1.6% a year, compared with 0.5% in China (Table 6.1). The present mean population density figure is 389 persons per km² compared with 253 for the UK and 140 for China. However, none of these densities comes anywhere near the 1,199 persons per km² recorded in Bangladesh.

Table 6.1 gives information about the current ten most populated countries in the world (the UK ranks number 20). The size of the populations of China and India is emphasised by the huge gap between them and the third-ranking country, the USA. Together the ten countries account for just under 40% of the world's land area and around 60% of its population. In terms of distribution, seven of the countries are located in Asia (the whole Indian subcontinent is included), two in the Americas and one in Africa.

Table 6.1
The ten most populated countries in the world, 2010

Rank	Country	Population, 2010 (millions)	Density, 2010 (people per km²)	Mean annual population change, 2000–10 (%)	HDI, 2007	Ecological footprint, 2000
1	China	1,330	140	0.5	0.772	1.54
2	India	1,173	389	1.6	0.612	0.77
3	USA	310	34	1.7	0.956	9.70
4	Indonesia	243	133	1.5	0.734	1.13
5	Brazil	201	24	2.1	0.813	2.38
6	Pakistan	184	227	2.9	0.572	0.64
7	Bangladesh	156	1,199	2.1	0.543	0.53
8	Nigeria	152	164	2.4	0.511	1.33
9	Russia	139	9	−0.4	0.817	4.55
10	Japan	126	349	−0.01	0.960	4.77

With respect to consumption, it is significant that two of the countries (the USA and Japan) are developed countries with typically high levels of resource consumption, which is reflected in their high ecological footprint values (see Figure 6.3 on page 74). In Japan, a stagnant or slightly decreasing population may help to ease the situation. However, the much higher population growth rate in the USA, with one of the heaviest ecological footprints, is worrying, as are the even higher rates in Pakistan and Brazil. It is also noteworthy that the table contains the four BRIC countries. Three of them (Brazil, Russia and China) already have deep ecological footprints and these will no doubt become even deeper as their economies expand and become more prosperous.

In 2050 it is estimated that the list of the top ten most populated countries will be somewhat different. India will top the table with an estimated population of 1.5 billion, followed by China (1.4 billion). There will be two newcomers, Ethiopia (169 million) and Congo (160 million) at 9 and 10 in the rankings. These will have replaced Russia and Japan. Clearly, the pressure on resources will become most critical in Asia (with five of the top ten countries) and Africa (with three of them).

Increased population may not necessarily mean ecological crisis, but it will not ease the pressure on already-stressed ecosystems. It is bound to create competition for, if not conflict over, scarce resources such as fresh water, food, energy, minerals and timber.

There is double trouble in store. The rich nations of the world, with their high consumption levels, have a record of polluting and destabilising the world far more than the poor. This is unlikely to change. Equally, land degradation and desertification are real and growing problems in developing countries, many of which are too poor to be able to cope with the challenges. Children born into at least half of these countries can look forward to a rough ride in life. Unless there is major change in the world, most of these children will live in cities and in poverty. There will be little or no narrowing of the range in **human development index (HDI)** values. There will also be widespread food shortages, as well as serious sanitation and health problems.

14 Using case studies

Question

Study Figure 6.6, which shows three different relationships between population and resources.

With reference to examples, evaluate the range of strategies available for countries to achieve the goal of an optimum population.

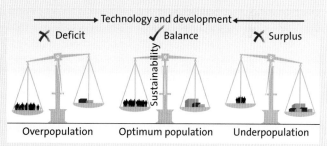

Figure 6.6 *Three different relationships between population and resources*

Guidance

The diagram suggests that there are three main courses of action:

- control population size (i.e. reduce overpopulation)
- control resource use (i.e. reduce consumption)
- use technology and development to bring population and resources into a sustainable balance

To reduce overpopulation, the main strategy is to target and reduce birth rates. Appropriate examples to use are China (reasonably successful — *Case study 29*) and India (distinctly unsuccessful). Another possible strategy is to change the distribution of population by encouraging people to move from congested to underpopulated countries or areas (Indonesia).

To reduce consumption, the aim is to steer away as much as possible from the use of non-renewable resources and to encourage recycling and conservation (the Netherlands is a leading light in this respect).

The trouble with both technological progress and development is that they have an in-built tendency to raise the consumption of resources There are few, if any, countries that have as yet really harnessed new technology to help reduce resource consumption and yet allow living standards to rise. Chile (*Case study 23*) is one country where the chances of achieving this sustainable balance are good, provided that there is a planned redistribution of the population and an investment of new technology in the opening up of regions that are presently underdeveloped.

Outlook

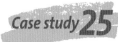
THE POPULATION TIME BOMB

In 2010, the world's population was fast approaching the 7 billion mark. Fifty years earlier, there were half that number of people, and in 1900 there were a quarter. Population has been increasing, largely as a result of the control of mortality through the medium of healthcare. This came about as a result of improvements in:

■ medicine and treatment of disease
■ access to healthcare
■ sanitation and personal hygiene
■ diet and food supply

As indicated in *Case study 5*, it now looks as if the rate of global population growth peaked in the 1960s at 2.0% per annum. It now stands at 1.3%. The rate may be falling, but because it is effectively a compound rate, large increases in population continue to occur each year. Forecasts indicate that even if the current trend in the growth rate continues, global population could reach 9 billion by 2050. Putting a firm brake on future population growth is going to require an effective programme of fertility control. This would require:

■ better availability and greater use of contraception
■ more effective education that emphasises the merits of smaller families and the responsibility of everyone to keep fertility in check
■ improved opportunities and choices for women

So it does look as if the global population bomb is beginning to be defused. However, burning questions remain:

■ Can the world cope even with its present level of population?
■ More critically, will it be able to cope with the population growth that will accrue between now and the day when zero population growth is achieved?

As for the first question, increasing poverty, malnutrition and starvation are indicators that even a population approaching 7 billion may be unsustainable. Those signs are most evident in the developing countries. Of the 5 billion people in developing countries, nearly three-fifths lack basic sanitation, about a third have no access to clean water, a quarter do not have adequate housing and a fifth have no access to modern health services. The daily calorie intake of a significant percentage is below that required to complete a day's manual work. However, the immediate problem is that at present something approaching 90% of babies born today will be raised in developing countries. Table 6.2 shows a major shift of the world's demographic centre of gravity towards the developing nations and to Africa in particular.

As for the second question, the outlook between now and the day of zero population growth is truly

Table 6.2
The changing distribution of the global population, 1950–2050

	% of world population		
	1950	2000	2050
Developed countries	32	20	13
Developing countries	68	80	87
Africa	9	12	20
Asia	55	61	59
Europe	22	12	7
South America and Caribbean	6	9	9
North America	7	5	4
Oceania	0.5	0.5	0.5

bleak. It is likely to be one of unbearable strains on resources and ever-deepening ecological footprints. Added to this will be an even sharper division of the world into:

- the 'haves' with their stagnant or even declining populations, their relative affluence and their high consumption of global resources
- the 'have nots' with their burgeoning populations, their acute poverty and their restricted access to global resources

It is the opinion of some that the population bomb has already 'exploded' and that these two global geographies — one of plenty enjoyed by a privileged minority, the other of hunger suffered by the majority of the world's population — are a major part of the fallout. Furthermore, these same observers suggest that the chances of ever achieving a sustainable global population have already been shattered.

A population approaching 7 billion is made unsustainable largely by a fundamental mismatch between the distributions of population and population growth on the one hand, and the distributions of food production and economic growth on the other. The chances of ever harmonising these distributions are very slim. But even if this were to happen to some degree, it might well be that the population bomb has already 'exploded' while it was being defused, splitting the world into two and destroying for ever the chances of attaining a sustainable global population.

Population policies

Policies refer to those actions of governments that try to control, manage or influence, in this instance, specific aspects of population and migration. To a lesser extent, decision makers in the private sector may also have some impact. For example, industrialists, in their recruitment of labour, may well encourage migration; so too developers, particularly through housing developments. However, the hand of public sector or government intervention can be seen in all of the four concepts covered so far (Figure 7.1). Governments seek to manipulate:

- the distribution of population
- population change
- population migration
- consumption, by actively encouraging the development process or more resource use

Six case studies follow to illustrate the first three of these policy 'areas'.

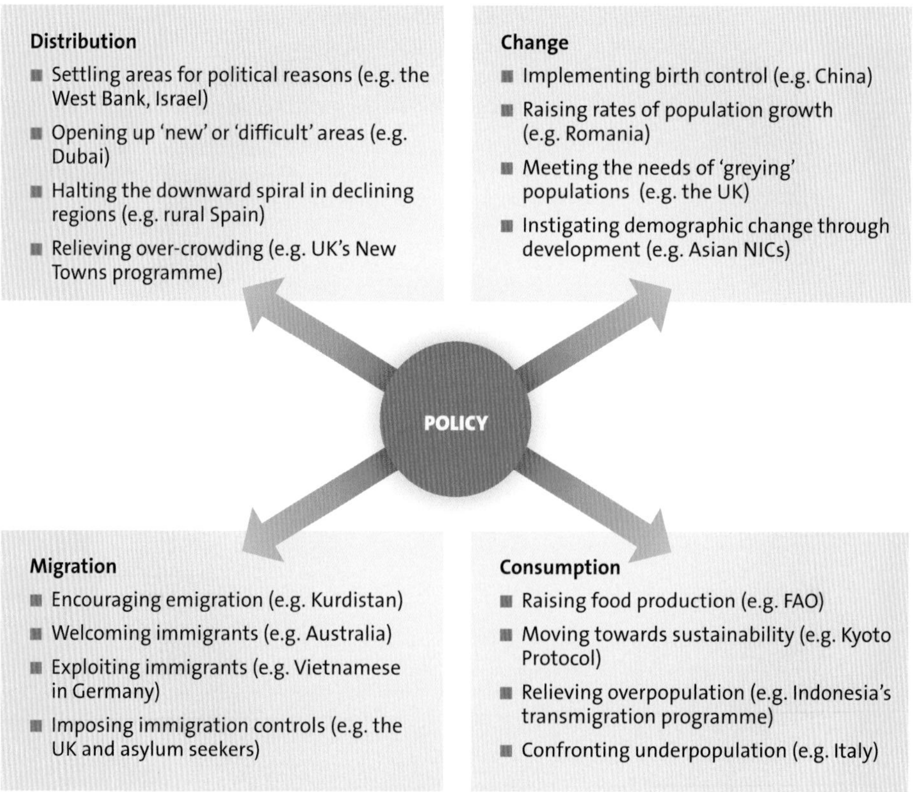

Figure 7.1
Some examples of population policies

Distribution
- Settling areas for political reasons (e.g. the West Bank, Israel)
- Opening up 'new' or 'difficult' areas (e.g. Dubai)
- Halting the downward spiral in declining regions (e.g. rural Spain)
- Relieving over-crowding (e.g. UK's New Towns programme)

Change
- Implementing birth control (e.g. China)
- Raising rates of population growth (e.g. Romania)
- Meeting the needs of 'greying' populations (e.g. the UK)
- Instigating demographic change through development (e.g. Asian NICs)

POLICY

Migration
- Encouraging emigration (e.g. Kurdistan)
- Welcoming immigrants (e.g. Australia)
- Exploiting immigrants (e.g. Vietnamese in Germany)
- Imposing immigration controls (e.g. the UK and asylum seekers)

Consumption
- Raising food production (e.g. FAO)
- Moving towards sustainability (e.g. Kyoto Protocol)
- Relieving overpopulation (e.g. Indonesia's transmigration programme)
- Confronting underpopulation (e.g. Italy)

Distribution

There are a number of different circumstances that might encourage governments actively to alter the distribution of population. These range from trying to achieve a spread of population that matches the distribution of resources and economic opportunities, to extending the nation's territorial grip.

JEWISH SETTLEMENTS IN THE WEST BANK: A POLITICAL POLICY

Case study 26

The state of Israel came into being in 1948, when the UN decided to split what had been Palestine and to make a Jewish and an Arab state. Needless to say, the Arabs resisted the partition and have continued to do so ever since. This part of the Middle East is one

Figure 7.2
Established Jewish settlements in the West Bank and Gaza

of the world's most persistent and violent trouble spots. To add to the tension, in 1967 Israel occupied a sizeable tract of the much-reduced Palestine (the so-called West Bank, roughly the size of Norfolk), mainly on the grounds that it would make Israel's borders more secure against attacks from its Arab neighbours. Since then, successive Israeli governments have pursued a policy of consolidating the West Bank as an integral part of Israel.

Three related strategies have been involved in the implementation of this policy. First, the Israeli government has sought to 'persuade' some of the Arabs to move away and settle in what is left of Palestine. Second, it has tried to concentrate the remaining Arabs in designated settlements and refugee camps. Recently, a British ambassador to Israel described the West Bank as 'the largest detention camp in the world'. Third, between the scattered areas of Arab settlement, the Israeli government has encouraged the building of well over 100 new settlements occupied exclusively by Israelis (many of whom are recent émigrés from eastern Europe) (Figure 7.2). The construction of some of these has involved the demolition of Arab settlements. Some of the new settlements have not been officially sanctioned.

The net outcome of these actions or policies has been to:

- raise the population density
- expand the settlement network
- concentrate the Arab people into fragmented areas
- increase the Jewish component in the population

Today, the population of the West Bank is around 2.2 million, with the Jewish element amounting to only 17%. An added difficulty in moving towards a Jewish majority in the West Bank is the much higher fertility of the Arabs — 34 per 1,000, compared with 20 per 1,000 for the Israelis.

While Israel may feel more secure with the 'converted' West Bank as part of its territory, the continuing terrorist activity, particularly the suicide bombings and rocket shells, must remind the Israelis that their occupation and settlement of it may not be too secure. Indeed, Israel has intermittently indicated a willingness to discuss the permanent status of the West Bank and Gaza (a strip of territory taken from Egypt in 1956) and to dismantle some of the 'illegal' Jewish outposts. However, so long as Jewish settlement continues, the hopes of a settled future for the West Bank remain remote.

This case study provides an illustration of how government policy can affect both the distribution and the composition of an area's population. In this instance, the reasons for doing so have been political and mainly to do with security. It remains in doubt whether the policy will ever achieve a permanent solution.

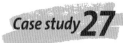 **Case study 27** ## DUBAI: PEOPLING AND GREENING THE DESERT

Dubai is one of seven small states, fronting the Arabian Gulf, that make up the United Arab Emirates (UAE). One hundred and fifty years ago, its territory of 3,900 km^2 was little more than barren, sandy desert. The only settlement was a small port of 1,000 or so inhabitants, located on the banks of Dubai Creek, and relying mainly on fishing, the pearl industry and overseas trade. Today, Dubai is an amazingly prosperous state. In addition to oil revenues, it thrives on a mix of corporate business (attracted there by zero taxation), re-exporting (helped by low duties) and tourism (thanks to its beautiful

philipus/Fotolia

beaches, warm seas, sumptuous hotels and duty-free shopping). The population has risen to nearly 1 million, most of whom live in the city itself. Dubai city is essentially linear, running parallel to the coast from Deira (old Dubai) in the direction of Abu Dhabi (another emirate). The built-up area has some of the most spectacular architecture in the world, notably the futuristic skyscraper office blocks and hotels, as well as many modern shopping malls (Figure 7.3).

Figure 7.3
Some of Dubai's dramatic architecture

It has been said that 'the story of Dubai reads like a rags-to-riches tale'. It is hard to imagine anywhere else in the world that has developed at such a pace, in such a short time. The desert sands are being hidden not just by the built-up area, but by grass and palm trees. One might say that the desert is beginning to 'blossom'. Much of Dubai's success has to be credited to the remarkable vision, enterprise and leadership of the ruling Maktoum family, who have been the policy and decision makers.

Migrant workers

The building and running of modern Dubai has involved, and continues to involve, two distinct waves of immigration. Construction has required the recruitment of technicians and professionals from western Europe and North America, and also of manual labour, mainly from Africa and the Indian subcontinent. The latter regions also supply most of the labour required now to fill a wide range of jobs, from shop-counter and hotel posts to domestic servants for wealthy Arab families and the general running of public services. This heavy reliance on immigrants is reflected in the composition of Dubai's population today. Only 18% of the population is made up of UAE nationals; 13% come from neighbouring Arab states, while 65% are expatriates from other parts of Europe and Asia. In order to gain entry into the country, an immigrant worker must be sponsored. Contracts of employment are for up to 2 years. However, an immigrant may only work for the sponsoring employer. For these reasons, there is a remarkable turn-over in Dubai's population. Sadly, most of the Asian and African immigrant workers are treated by the Dubaiites as second-class citizens. If a worker earns less than 3,000 dirhams (approximately £500) a month, they are not permitted to bring their spouse and family into the country.

Resource consumption

If one had to name one resource, other than oil, that has been crucial to the rise of modern Dubai, it would be water. In a virtually rain-free part of the world, Dubai has had to look to the desalination of seawater to provide most of the 700,000 m³ of water needed each day. Apart from human consumption, water is needed for the constant irrigation of golf courses, grass verges and palm trees. Water lost through high rates of evaporation from the many fountains that punctuate the built-up area also needs to be replaced. It is small wonder that Dubai has been at the cutting edge of new desalination technology. It has pioneered the building of some of the largest desalination stations in the world.

Resource consumption in Dubai involves much more than just oil and seawater. The extravagance of nearly 1 million inhabitants and over 3 million tourists a year is immense. The demands range from electricity (much of it to help Dubai keep cool in sweltering temperatures) and high-grade building materials to top-quality food and the latest in electronic gadgetry. Dubai's ecological footprint is indeed a deep one.

This case study makes the basic point that even the most hostile of environments can be settled in and developed. Success or failure depends largely on:
- *sound decision making and management*
- *the use of modern technology*
- *the availability of large amounts of capital*

15 **Using case studies**

Question

'The impact of the physical environment on the distribution of population is greater in developing countries than in developed countries.' Discuss this statement.

Guidance

An answer to this question might make use of *Case studies 9* and *10*. Since the latter may be argued as an exception to the rule, it would raise the overall quality of the answer if we were able to introduce another developed country example, but one that conforms. You could choose *Case study 26* or even *Case study 27* for this purpose.

You might start by pointing out that the chosen countries differ not just in terms of development but also in terms of their physical environments. Japan covers a large latitudinal range, from cool temperate to subtropical; three-quarters of it is inaccessible upland. The relatively small amount of lowland is highly fragmented and occurs mainly around the coast. By contrast, Ethiopia shows great altitudinal range, which covers roughly the same range of climates as in Japan. Israel is a small country. Extensive tracts of arid and semi-arid terrain lie between the coastal belt in the west and the River Jordan in the east.

Having described the physical background, we should now move on to the patterns of population distribution. For all Japan's economic strength and technological know-how, the distribution of its population is very much governed by the distribution of lowlands. Much of the upland interior of the country is inaccessible and of little value except for its forests and for some forms of tourism and recreation. In Ethiopia, with its poverty and little know-how, the population tends to avoid the extreme altitudes (because they are either too cold or too hot and arid). In Israel, much has been done to 'green' the arid areas and to reduce the early concentration of settlements along the coast and Jordan valley.

Thus, in two of the countries, despite the huge development gap between them, the physical environment exerts a potent influence on the distribution of population. Japan refutes the generalisation, but Israel supports it in that modern technology has been used to overcome a distinctly hostile physical environment and to even out the distribution of population.

Change

While the rate of population growth has for a long time been a major concern, recent history is peppered with examples of governments wishing to boost their populations. This was the case in Europe after the First and Second World Wars. There was a desperate need to make up for those killed during the fighting and to restore the labour force to the required level. More recently, some developed countries have become alarmed by the rapid 'greying' of their populations (*Case study 33*) and by the depopulation of rural areas (*Case study 31*). Basically, there are two ways to boost a population. One is to encourage couples to have more children. The second is to encourage immigration, particularly of people in the reproductive age range. The latter option offers a 'quicker fix' and seems to be more favoured at the present time, as illustrated by Spain (*Case study 31*).

After the Second World War, France and the UK took both options, but they have since gone rather cool on immigration (*Case studies 13, 20 and 35*). Perhaps the most draconian example of taking the first option was provided by Romania during the period 1966–90. Actions included taxing childless couples, banning the sale of contraceptives, paying generous benefits to families and lowering the legal age of marriage to 15.

As more countries enter Stage 5 of the DTM, so governments are being forced to face the challenge of how best to save their countries from demographic implosion.

Countries reaching what appears to be Stage 5 in the demographic transition model (DTM) are being confronted by shrinking populations and underpopulation. The new scenario is well illustrated by two countries on opposite sides of the world — Japan and Italy.

Japan

Population change in Japan has now passed below the replacement level, and the birth rate is about to pass beneath the death rate (Figure 7.4). Once the latter happens, the total population will begin to fall from its present peak of 127 million. The combination of a shrinking population and an increasingly elderly one will have a growing impact on everything from demand for goods and services to public finances and the structure of the labour force. For example, within 5 years, Japanese universities expect to have more available places than applicants. The effects on the workforce could be dramatic. The number of people in the economically active age range has been falling since 1995. As the working cohort shrinks to something less than half the population by 2050, so the number of elderly people will rise to account for one-third.

From the point of view of the economy, Japan is fast becoming underpopulated. It is running out of people to keep the economy turning over at its present level. During the next 10 years, the labour shortfall is expected to rise to 5 million.

Japan is also going to become underpopulated in terms of the vast investment that has been made in the past to provide a range of services for a younger population. Schools, colleges and universities are one such category of service. More and more of the infrastructure is becoming surplus to requirements. On the other hand, delivering all that is required by its 'greying' population cannot be guaranteed when the economy is spluttering through lack of workers. So, perhaps we begin to see a paradox. Japan is underpopulated in terms of the economy and young people, but it is clearly becoming overpopulated by the elderly.

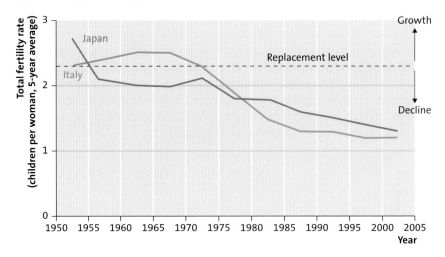

Figure 7.4
Japan and Italy: changing total fertility rates

The Japanese government has yet to formulate a policy to cope with this emerging scenario. The attitude seems to be almost complacent — a serious demographic problem is approaching fast. The country really has only two options:

- Offer massive incentives to the remaining young adults to have more than two children. A major drawback here is that dwelling units are both small and very expensive due to the very limited amount of habitable space.
- Meet the labour shortfall by attracting immigrant workers. This is not a popular idea with many Japanese people, as they like to think of themselves as an ethnically homogenous nation. But the harsh reality is that this may now be the only way of swelling the country's population of reproductive age and meeting the growing labour shortage.

Italy

The demographic situation in Italy is very similar to that of Japan. Italy has one of the lowest levels of fertility in the world. The total fertility rate is 1.23 children per woman, which is far below the replacement level of 2.3 children (Figure 7.4). Demographers calculate that by 2050 the current population of 57 million could have dwindled to 41 million. If so, towns and cities could be left with thousands of unwanted apartments, some schools may well be half empty and large tracts of the countryside could be depopulated. By 2050, there could be one pensioner for every one productive worker. It raises the interesting question: how is Italy going to support so many pensioners?

In 2003, Italy introduced the 'baby bonus' of €1,000 for mothers who have a second child. This has now been extended to each first child. Some local authorities have gone further and launched 'baby-making' schemes of their own. For example, in one locality couples are being offered €10,000 for every baby born. However, it is unlikely that incentives such as these can really change the population 'landscape'.

It is perhaps worth asking the question: why does Italy not go for the UK solution of mildly encouraging immigration? After all, there are thousands of Albanians just across the water simply itching for a better life. What do you think of this possible solution?

Figure 7.5
Total fertility rates and population prospects, 2005

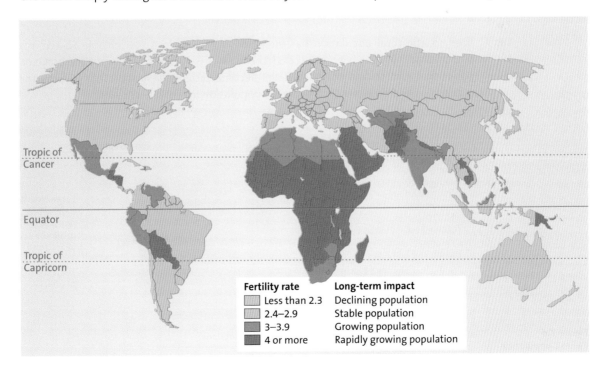

Fertility rate	Long-term impact
Less than 2.3	Declining population
2.4–2.9	Stable population
3–3.9	Growing population
4 or more	Rapidly growing population

Population & Migration

Figure 7.5 shows that a declining population is not just Italy's problem — it is Europe's too. As yet, none of the countries sharing the problem has policies in place to confront the situation. As one expert has put it: 'Europeans will probably not wake up to the problems until they are all in their wheelchairs and then suddenly realise there is no one to push!' Is this the endgame to this 'new' form of underpopulation?

Rather than being about specific policies, this case study shows that governments can be slow to respond to situations that are clearly going to happen. What are the reasons — politics, complacency or something else? You might also extend coverage of this topic by looking at Case study 31. It is ironic that while the world as a whole faces a population 'explosion' (see Case study 25) there are countries facing a population implosion.

Although there is a growing band of countries facing serious declines in their populations (negative demographic dividend), rather more governments have been confronted with the opposite challenge of how to curb high rates of population growth (positive demographic dividend). In terms of positive and effective anti-natal action, there is still none to rival China's one-child policy.

Case study 29 — BACKLASH TO CHINA'S ONE-CHILD POLICY

Much has been written about China's attempts to curb the rate of growth of its immense population. In 1980 it introduced the draconian one-child policy. Except in a few particular circumstances, for nearly 20 years no couple was allowed to have more than one child. Those who did were penalised in various ways. Since 1999, there has been some relaxation of the policy, particularly in rural areas. However, the impacts of the original policy are evident in:

- a reduced birth rate (from 31 to 17 per 1,000)
- a reduced growth rate (from 2.4% to 0.9% per year)

Even so, the total population has increased from 996 to 1,266 million. So has the one-child policy really been a success? The answer is partly so. It has put the brake on population growth and that brake will become progressively more effective as the cutback in children works its way up the population pyramid (Figure 7.6).

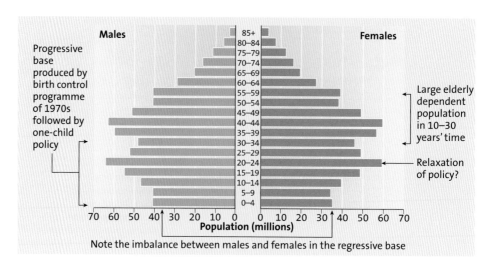

Figure 7.6
China's population pyramid, 2010

Note the imbalance between males and females in the regressive base

However, a number of unforeseen demographic backlashes are already apparent:

- **uneven implementation** — the policy seems to have been implemented more effectively in towns and cities than in rural areas
- **negative growth** is now being experienced is some cities (e.g. Shanghai), where the policy may have been too effective
- **sex-selective abortion** — the traditional wish of fathers to produce a son remains strong; the gender ratio now stands at 116 males to every 100 females
- **increased divorce rate** — divorce is used by men as a means of ensuring a male heir
- **'little emperor' syndrome** — parents 'spoil' their 'one boy' child, who as a result may become obese, demanding and delinquent
- **shift in dependency** — while having fewer children lowers dependency at one end of the age spectrum, it raises it at the other end; the bulge in the middle of the present population pyramid (Figure 7.6) will soon result in more old people being supported by a reduced working population
- **gender imbalance** — created by the preference for a male child, this promises some serious social problems. For example, due to the increasing shortage of women of marrying age, bride bartering or kidnapping has already become commonplace in rural areas. Prostitution and the trade in sex slaves are now rife in the cities. The growing population of unattached 'little emperors' (Figure 7.7), particularly in the cities, could easily spark social instability and a rise in crime. A possible solution might be to encourage the immigration of women from other countries.

Figure 7.7
Chinese children in the classroom. How many of the boys will be able to marry when they are older?

The bottom-line question on this issue of gender imbalance is: why do Chinese parents continue to show a strong preference for boys over girls? In India, the explanation lies in the dowry tradition, but that tradition does not exist in China. Instead, it looks as if the reason may lie in the absence of any pension system and the belief that it is sons, not daughters, who can support parents in their twilight years. In 2004, the government recognised the problem and introduced its 'Care for Girls' plan. Farming families are to be offered incentives to stop the abortion of female foetuses. Girls will be exempted from school fees, while their parents will benefit from tax breaks, free insurance and extra housing.

The future

Further down the line, it looks as if labour shortages will be another negative consequence of the negative demographic dividend. At the moment, unemployment prevails in most cities, a problem made worse by large-scale rural-to-urban migration. However, that will change as the reduced cohorts of children become part of the productive population. Although not directly linked to the one-child policy, it is worth noting that the food security of China is being threatened, as urban areas expand onto scarce arable land.

Even further down the line (say, in 50 years), it is possible that, despite a relaxation of the policy, China's population might contract to 800 million. The population pyramid will then look like those of developed countries that are entering Stage 5 of the demographic transition model. Is that what the policy makers in China really want?

Late in 2010, family planning authorities in China were considering the most sweeping liberalisation yet of their strict birth-control policies. A pilot project to be trialled in several provinces would allow couples where one partner is an only child to have two children. This relaxation is being driven by official concern about growing shortages of labour across much of the country and the rapid ageing of the country's population structure.

This tough policy has done much to defuse the population explosion that so threatened China in the third quarter of the twentieth century. However, the case study also illustrates two vital points:

- *A policy aimed at one objective can often have unwanted and unforeseen impacts.*
- *A policy that prevails for too long can become counterproductive — such is the dynamic nature of all populations.*

Using case studies

16

Question

Assess the extent to which the following factors are crucial to a successful birth control programme:

- **accessible and affordable contraception**
- **female empowerment**
- **education**
- **religion**
- **economic incentives**
- **political persuasion**

Guidance

Answering this question requires explaining how each factor impacts on birth control. For a high-grade response, you will need to undertake some evaluation: that is, draw attention to what you believe are the most significant factors.

Migration

It is no exaggeration to say that international migration is a major global issue. Increasingly, governments at both ends of the migration pathway feel the need to intervene, mainly to put the brake on either outflows from major source countries or inflows to popular host countries, particularly the latter. Migration policies often have sinister political motives. Equally, many governments are finding it increasingly difficult to curb the forces of economic globalisation that are so encouraging to international migration. The following case studies are somewhat exceptional in that they deal with two countries with distinctly positive attitudes towards immigration.

POPULATING AUSTRALIA THROUGH IMMIGRATION

Although Australia had, and still has, an aboriginal population (currently estimated at around 0.5 million), much of the peopling of this vast country has been through immigration. Initially, migration policy as such was in the hands of the British government. In 1788 Australia (strictly speaking New South Wales) had been acquired as a British colony (Figure 7.8). The main reason was that, after the loss of the USA, Britain needed a new penal colony for the relocation of convicts from its overcrowded prisons. Between 1788 and 1868 (when the penal function was brought to an end), around 160,000 men and women were deported to Australia. Many of those prisoners who survived to serve out their sentences opted to remain in the country. In 1815, the British government began to encourage the emigration to Australia of 'free settlers'. This policy was to continue for nearly 150 years. Initially, many people were attracted by the prospect of beginning a new life on free Crown land. After the Second World War (1945), there was the 'assisted passages' scheme in which UK citizens could emigrate to Australia from a mere £10.

Figure 7.8
Australia: an immigration timeline

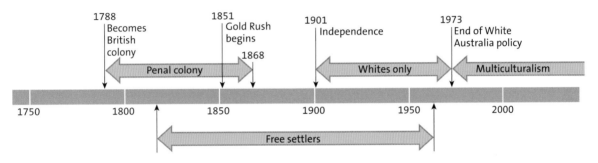

The discovery of gold in 1851 and the ensuing 'gold rush' boosted the peopling of Australia. Thousands, mainly drawn from the British Isles and parts of Europe, entered the country in the hope of making their fortunes.

'Whites only'

In 1901 Australia was granted virtual independence and ceased to be a British colony. It is interesting to note that one of the first pieces of legislation passed by the new Australian government was the 'White Australia' policy. Fearing that the country might be overrun by immigrants of Chinese origin, only settlers of European origin were to be admitted.

Up until the end of the UK's assisted passage scheme in the 1960s and Australia's 'whites only' immigration policy (1972), most new settlers came from the UK. This is still reflected in the fact that roughly a quarter of all of today's 'overseas-born' Australians started their lives in the UK. While the figure of around a quarter also applies to the present inward flow of migrants, it conceals the fact that there has been a fundamental shift in the 'sourcing' of settlers. Today, nearly half come from

Figure 7.9
Australia: overseas-born population by region, 2006

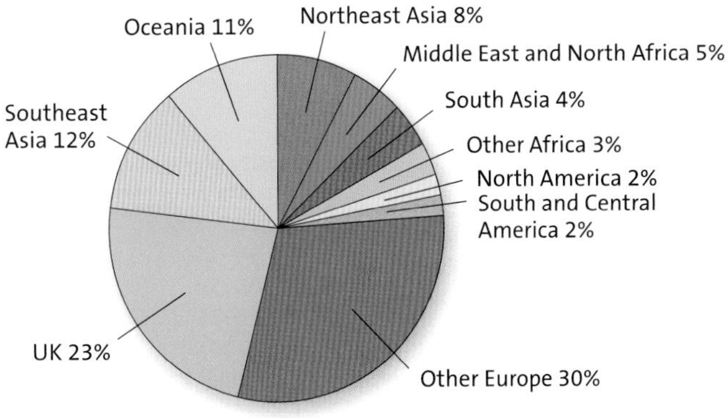

Oceania 11%
Northeast Asia 8%
Middle East and North Africa 5%
South Asia 4%
Southeast Asia 12%
Other Africa 3%
North America 2%
South and Central America 2%
UK 23%
Other Europe 30%

outside Europe, and half of these are drawn from various parts of Asia (Figure 7.9). The change in immigration policy, from 'whites only' to what is referred to as 'multiculturalism', was mainly motivated by the government's wish to boost the population of this vast country. In particular, it wanted to encourage the settlement of the country's physically hostile 'dead' or empty heart (Figure 7.10). The government was also keen to raise the country's birth rate. This had been running at a persistently low level because of typically low fertility rates among European settlers.

The last 30 years

For much of the time since 1981, natural increase has been running at a little over 30,000 per annum, but in the last few years it has risen to 40,000 (Figure 7.11). During the same period, net overseas migration has fluctuated, but in 2009 it reached a record level of 70,000. Immigration remains carefully controlled by government. While present policy is much less selective in terms of ethnicity, it is now largely based on the skills and age of would-be immigrants. The issuing of immigration visas is now governed by a points-scoring system based primarily on a prioritised list of the occupational skills that Australia needs.

Figure 7.10
Australia: distribution of population, 2001

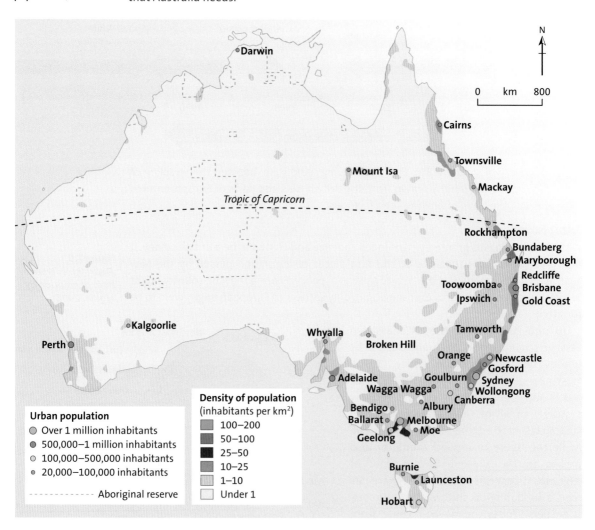

Contemporary Case Studies

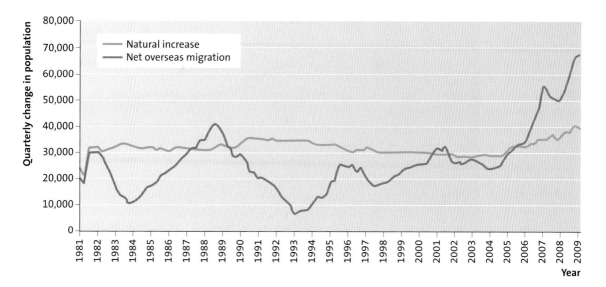

Figure 7.11
Australia's components of population growth: June 1981–June 2009

It should also be noted that Australia has a humanitarian programme, which accepts up to 15,000 refugees each year from the world's trouble spots. For a while, many of these refugees came from Africa, particularly Sudan. However, because of integration problems, in 2007 a ban was placed on admitting any more African refugees. Instead, priority is currently given to refugees from Myanmar and Afghanistan.

What is interesting is that the shift in the sourcing of immigrants from Europe to Asia does not appear, as yet, to have had much impact on the birth rate, which continues to fall from a high of 23 per 1,000 in the 1950s. It seems that the perception in developed countries that the smaller the family, the better is the quality of life has caught on among the new migrants. The birth rate currently stands at 14 per 1,000 and the country's population now totals over 20 million.

Figure 7.11 shows that net migration has been an important component of population growth. But while immigration has been successfully managed by successive Australian governments, little progress has been made in terms of peopling the country's vast interior. Recently, in an attempt to overcome this deficiency, the Australian government introduced a scheme whereby virtually anyone is allowed to enter the country provided they spend their first 10 years living in the outback. It remains to be seen what impact this might have on the distribution of population.

The bottom line to this case study is that, while Australia has been successful in attracting multi-ethnic immigrants to help meet the huge challenge of populating this vast country, there has been no substantial alteration of the traditional population distribution pattern. The physical hostility of the country's desert interior and the momentum of settlement along the coast have resulted in Australia retaining its 'dead heart'.

REPOPULATING RURAL SPAIN

Case study **31**

The migration history of Spain shows two major U-turns. For centuries, up to the last quarter of the twentieth century, Spanish migration was about emigration. Generations of Spaniards struggling to make a living perceived the Americas as the 'promised land'. However, after the Second World War the migration pattern changed completely, with

Figure 7.12
A village in Teruel Province of Spain, where lack of population has led to decline

Antoni Traver/Fotolia

three-quarters of emigrants heading for the countries of northern Europe. Up to 100,000 Spaniards were leaving every year.

Moving in not out

An even more profound shift in the pattern of migration occurred after 1970, when immigration became the dominant flow. In 2000, for example, 90,000 new immigrants arrived, while emigration dried up to a trickle of 2,000 people. The reason for the U-turn was the take-off of the Spanish economy. Having exported its labour force for so long, Spain suddenly found itself desperately short of workers. A low and dropping birth rate meant that natural increase would not meet the shortfall. The only solution was to import cheap labour, mainly from Morocco and other parts of Africa that had once been Spanish colonies. Those described as 'foreign residents' now account for nearly 3% of Spain's population.

While Spain has had no option but to let in foreign workers, it has struggled to keep immigration under control. Its wish has been to regulate the in-flow to ensure that immigrant skills closely match those required by the economy. In the event, the Spanish authorities have become concerned about the nature of the immigrants. In their view, too many have been coming from North Africa and they are often ill suited to the specific needs of the service sector vacancies. The authorities have also become alarmed by the rising tide of illegal immigrants. The Spanish have discovered what many other countries have before: the more you tighten up on immigration, the more this seems to encourage illegal entry (see Figure 4.4 on page 47).

'Returning' to Spain

A dramatic turn of events occurred in January 2003 when the Spanish government passed a law encouraging various categories of 'non-citizen' to apply for passports. These 'non-citizens' are mainly the descendants of two groups of Spanish emigrants: those who settled in Latin America and those who moved to other European countries as refugees to escape the fascist regime of General Franco (a dictator who ruled Spain from 1936 to shortly before his death in 1975). Both groups were suddenly deemed to

Contemporary Case Studies

be Spanish. The upshot is that around 1 million people may have the right to return to Spain — 850,000 from Latin America and 150,000 from Europe. Spanish embassies in Latin American countries have been inundated by tens of thousands of applicants for Spanish passports. The biggest queues have been in Argentina, a country torn apart by economic crisis and home to the largest overseas Spanish community.

This sudden U-turn has given momentum to an initiative taken a few years earlier by some local authorities in Spain to halt rural depopulation. Since the 1950s, about 2,000 remote settlements across Spain have been abandoned as a result of emigration and rural-to-urban migration. However, the tide may be beginning to turn in the remote mountainous area of Teruel Province to the east of Madrid. This is the least populated part of Spain, one of the poorest and the only one where deaths exceed births. For several years now, village authorities have been offering Latin Americans free plane tickets, jobs and houses to encourage them to move to this isolated area, in a last desperate attempt to stop their settlements disappearing altogether (Figure 7.12). Settlers are also being sought in eastern Europe.

The leader in this search for new blood has been the village of Aguaviva. In 2000, its mayor set up the Association of Spanish Villages against Depopulation. Membership has already risen to 130 communities. The association is looking for married couples under 40, without college degrees, with work permits (or one of the new passports) and at least two children under 12. Qualifying families sign a 5-year contract. In return, they are given jobs paying between £425 and £600 a month, housing at a monthly rent of £100, free healthcare and schooling. The available work is mainly in farming and services, but it is recognised that any lasting revival of the villages will need the creation of more jobs, particularly for women, the young and people over 45. Of course, it is hoped that after the 5 years are up, families will be happy to stay on.

Most of the recruiting for new settlers has taken place in Argentina and Uruguay, but Aguaviva now has 100 new residents from Romania. In this village, the effects of the scheme are already visible. The average age of the local people sitting in the village square's bar taking an afternoon drink is 82, while that of the incomers is 31.

Thus, centuries after Spain sent its sons and daughters to colonise the New World (the conquistadors), it is now trying to persuade their descendants to return and help to revive Spain's dying country villages.

This case study illustrates the 'swings and roundabouts' that mark one country's migration history. Change in the balance of push and pull factors has led to a complete reversal of the dominant migration direction. While migrants themselves are decision makers, the journeys they make are conditioned by the decisions and policies of governments, both national and local.

17 Question

Using case studies

Read *Case study 31* again and produce an annotated timeline diagram that summarises Spain's main migration trends since *c.*1900.

Guidance

Figures 5.5 and 7.8 might give you some ideas.

Part 8

Population issues

This part of the book looks at five highly topical issues. Four of them relate directly to one of three key concepts underlying the book: namely, population change (*Case studies 32* and *33*), population distribution (*Case study 34*) and migration (*Case study 35*). It would be an exaggeration to claim that these issues are of a truly global scale, but they are of consequence to significant but different sections of the global community. Two issues — the continuing scourge of HIV/AIDS (*Case study 32*) and megacity mania (*Case study 34*) — are particularly critical in developing countries, while the issues of greying populations (*Case study 33*) and freedom of movement (*Case study 35*) are of major concern in an increasing number of developed countries. The final issue (*Case study 36*) is truly global and serves as a sobering reminder that in so many different ways we live in a very unequal world. All five issues have major policy and management implications.

Case study 32 — THE CONTINUING SCOURGE OF HIV/AIDS

Three decades after AIDS was first identified, the pandemic has a deadly grip on the world. In 2008, an estimated 31.3 million people carried the HIV infection that causes AIDS. The annual death toll in that year was over 2 million. After years of denial that the disease existed, the developing countries (where most of the affected live) are gradually waking up to the need for action. More political energy is being focused on the challenge. Huge sums of money are being raised (nearly $9 billion in 2006) by a range of donors, from UN agencies and multilateral organisations to individual governments and private groups. This money and the political will are being directed at three priorities:

- teaching about healthy transmission-prevention techniques
- caring for the millions of children orphaned by AIDS
- treating those already infected

Sub-Saharan Africa

Sub-Saharan Africa accounts for more than 22.4 million or 72% of all the people with HIV/AIDS worldwide. It also accounts for 15 million of the estimated 20 million deaths caused by the disease. In 2008 around 1.4 million people died from AIDS in sub-Saharan Africa and 1.9 million people became infected with HIV. Since the beginning of the epidemic, more than 14 million children have lost one or both parents to HIV/AIDS.

   While these statistics are truly horrific, the situation is not universally bleak. There are some rays of hope, thanks to a handful of responsible governments. For example, the government of Senegal showed foresight by launching prevention programmes at a time when the country was largely untouched by the disease. This seems to be paying off — only 27,000 of its 9.6 million people are HIV-positive. Uganda is another 'beacon of hope'. In the early 1990s, the HIV prevalence rate reached 14%, with some urban areas suffering 31%. The national rate has now fallen to 6.1%. A concerted anti-HIV campaign, backed by the country's religious, traditional and civic leaders, bombarded Uganda's 23 million people with a very strong anti-HIV message. This resulted in a substantial increase in sexual abstinence among the young and an increase in condom use from about 7% to 50% in rural areas and to 85% in urban areas.

Compare these two encouraging examples with the situations in Swaziland, where the infection rate is 26.1%, and Botswana (23.9%). Almost as bad is the case of South Africa where, until quite recently, former President Mbeki was questioning whether HIV/AIDS existed at all! Yet South Africa has the largest HIV-positive population in the world (over 8 million in a population of 44 million). Roughly 40% of all adult deaths are now due to AIDS. By 2015, the disease is expected to have lowered the population by 12 million. Projections of present trends in Swaziland and Botswana are still more alarming. They raise the distinct possibility that their populations could become extinct in a matter of decades unless some decisive action is taken immediately.

Figure 8.1 illustrates how the spread of HIV/AIDS touches all five components recognised as part of population geography. However, the contact is a double one. Two distinct contexts of cause (contributing to) and effect (consequences) are involved. The consequences are first and foremost demographic. The diagram also hints at some of the courses of action needed to fight the spread of the pandemic. For example, education is critical to the success of any anti-AIDS campaign. The young need to be warned of the dangers of sexual promiscuity, unprotected sex and needle sharing among drug addicts. Transmission of the disease is particularly encouraged by the migratory labour systems that characterise much of southern Africa. These involve men leaving their homes and families to find work, sometimes abroad or in distant cities, often in the mining industry and transport. Time away from home often leads to time being spent in brothels. However, there are also customs and practices deeply embedded in African culture that put people at risk, including:

- polygamy (taking more than one wife)
- inheriting the wife of a deceased brother
- communal breast-feeding, which is still widely practised in rural areas

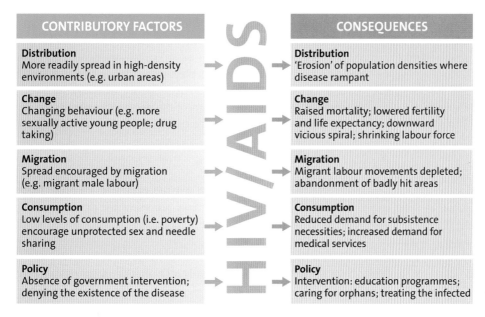

Figure 8.1
The spread of HIV/AIDS: contributory factors and consequences

It is beginning to be recognised that the spread of HIV/AIDS is a global issue:

■ There are few, if any, countries where the disease is not encountered.
■ The global community has a duty to help those countries most afflicted by AIDS. This obligation is heightened by the fact that many of those countries are poor. They simply do not have the resources and the organisational structures to halt the spread of the disease. Anti-retroviral drugs are now available; these need to be made freely available to sufferers, particularly the poor ones.
■ There is the argument that until HIV/AIDS is curbed in sub-Saharan Africa, there remains the risk of it spreading to and reinfecting other parts of the world.
■ The pandemic is worsening severely in countries that have so far escaped lightly, particularly the 'demographic giants' of India, China, Russia and Indonesia.

The potential demographic and wider consequences of the spread of HIV/AIDS are so horrifying as to be almost unthinkable. The only hope for the future lies in a combination of bottom-up and top-down strategies, ranging from more responsible individual behaviour and better education to governmental leadership and international aid.

The next case study may seem a little perverse, in that case studies elsewhere in the book, particularly in Part 6, have warned us of the need to curb the growth in global population. However, you should be saying to yourself: 'Aren't an increasing number of countries reaching Stages 4 and 5 of the DTM? Surely, this is only achieved if fertility falls to a permanently low pitch? Isn't this good news? What can the problem possibly be in such countries?' Problems and challenges there certainly are (see *Case study 28*), but a 'silvering' society also has its benefits.

Case study 33 — 'GREYING' POPULATIONS

In 2000 the UK was one of 61 countries in which not enough babies were being born to replace their populations. The UK's population was forecast to shrink to 56.7 million by 2050. But this predicted decline has been reversed due to a rise in both birth rate and immigration. Despite this, fertility remains so low in 22 European countries (see Table 8.1) and in HICs elsewhere that their populations will decline. As should now be well understood, decline leads inevitably to an ageing or 'greying' society. The age–sex pyramid is progressively undercut at its base and the upward taper becomes much blunter. The most obvious outcome of this is a shift in dependency, whereby the elderly outnumber the young. This in turn changes the nature of the demand for social services. Crudely put, demand shifts from crèches and schools to sheltered accommodation and hospital beds.

The ageing society is usually portrayed as a nightmare scenario of hospital beds and care homes packed with frail elderly, of mass poverty among growing ranks of pensioners and of fewer people to provide much-needed care and support. These things might happen and there is no denying that they represent the downside of an ageing society. However, they could be counteracted by making some socioeconomic adjustments, including:

■ raising the pension age to at least 65 (perhaps 67 or 70) for both men and women
■ lowering the high turnover in labour caused by early retirement and the recruitment of younger employees

- requiring that workers do more to set up their own pension funds, rather than relying on an increasingly inadequate state pension
- changing the mindset of employers to recognise the value of experience in the labour force

Action along these lines would undoubtedly maintain a larger labour force. It would also improve a country's ability to support growing numbers of elderly people.

Positives

Even without these changes, there is much to celebrate in greying societies — increasing life expectancy, longer retirement, more choice. A profile of 'ancient Britain' in 2000 included the following facts:

- There were 10.7 million people aged 50 or over.
- 5.4 million women were aged 65 or over compared with 3.9 million men.
- A man aged 60 could expect to live for another 19 years and a woman another 23 years.
- Of men aged over 65 and of women over 60, 8% were in employment.
- Since 1980, the numbers of over-50s in work had fallen by 20% and the numbers of over-60s by 30%.
- 70% of pensioners depended on state benefits for over 50% of their income.
- Of people aged over 75, 29% of men were widowed, compared with 61% of women.

A combination of good health, paid-up mortgages and 'empty nests' means that many people in their sixties and seventies are taking up new interests, going back to the classroom and fuelling a boom in the leisure market. A significant growth of overseas 'ski' ('spending the kids' inheritance') travel is perhaps spreading benefits to those developing countries currently viewed as attractive tourist destinations.

Table 8.1 *Getting smaller: Europe's national population decline, as predicted in 2000*

Country	Population, 2000 (thousands)	Population, 2050 (thousands)
Austria	8,142	7,094
Belgium	10,301	8,918
Bulgaria	7,707	5,673
Croatia	4,297	3,673
Czech Rep.	10,197	7,829
Denmark	5,375	4,793
Finland	5,187	4,898
Germany	81,895	73,303
Greece	10,559	8,233
Hungary	10,029	7,488
Italy	57,513	41,197
Lithuania	3,582	2,297
Netherlands	16,078	14,156
Poland	38,646	36,256
Portugal	10,084	8,137
Russian Fed.	143,300	121,256
Serbia-Montenegro	10,701	10,548
Slovakia	5,400	4,836
Spain	40,586	30,226
Sweden	8,924	8,661
Switzerland	7,319	6,745
Ukraine	49,927	39,302
UK	59,734	56,667

Social change

It is anticipated that popular culture will change. When over half the population is over 50, it will no longer be dominated by the obsession with youth. Society's institutions will have to adapt to the interests of older people. Most crimes are committed by the young, and the elderly are far more law abiding, so the crime rate is expected to fall steadily. There will be less need for police and prisons. Politics will become more stable and less prone to knee-jerk reactions and seesaws in policy. The 'grey' lobby will grow ever more powerful and difficult to ignore. Elderly people have more experience; they have seen it all before; they are more conservative. Elderly people are becoming more participative and proactive in politics, voluntary work and community associations.

The changes of a 'greying' society are likely to bring parents and children closer together, changing the relationship from one of dependency to one of equality. Longer life expectancy means that parent and child now enjoy a longer adult relationship free of the reciprocal dependency responsibilities of either child rearing or caring for the elderly in their last years.

At present, the scenario just painted applies only to a small, but growing, number of developed countries. One wonders how long it will be before it becomes a truly global issue: or might it be that in many of today's developing countries the HIV/AIDS pandemic will delay its onset indefinitely?

18

Using case studies

Question

Discuss the view that the challenge of a 'greying' population is now a global issue.

Guidance

Figure 8.2 demonstrates the spider-diagram technique that can be usefully applied in planning essays and answers to examination questions.

Having 'thought-bombed' (brainstormed) the key points you think need to be made (Figure 8.2), it is then necessary to order them into a coherent sequence. Alternatively, you might rank them on the basis of their importance to your discussion.

Try to number the boxes in Figure 8.2 and write an answer. Of course, you might wish to add more points to the diagram before you start writing.

The DTM suggests that most, if not all, countries will eventually reach at least Stage 4. Europe and North America today — where tomorrow? If true, certainly this will make 'greying' a global issue.

'Greying' societies mean HICs are less able to help LICs. This helps make the issue a global one.

'Greying' countries suffer increasing labour shortages. Who works the economy and who provides services, especially those for the elderly?

Immigration is a short-term solution to labour shortages. Presumably immigrants will come from current LICs. This helps make the issue a global one.

What happens when the labour shortage becomes a global one? World governments will have to promote pro-natalist policies largely by means of incentives.

'Greying' populations hold out the promise of a global future that is sustainable so far as the environment and resources are concerned.

POSSIBLE CONCLUSION

The issue is becoming a global one. It looks as if the world may be sitting on a demographic time bomb.

Figure 8.2
A spider diagram to use as an answer plan

Case study 34   MEGACITY MANIA

Just over half of the world's population now lives in an urban environment. One of the most spectacular features of recent urbanisation has been the emergence of huge metropolitan areas, widely referred to 'megacities'. Some set the population threshold of megacity status at a minimum of 10 million inhabitants, others at 5 million. The current distributions of urban areas passing both thresholds are shown on Figure 8.3.

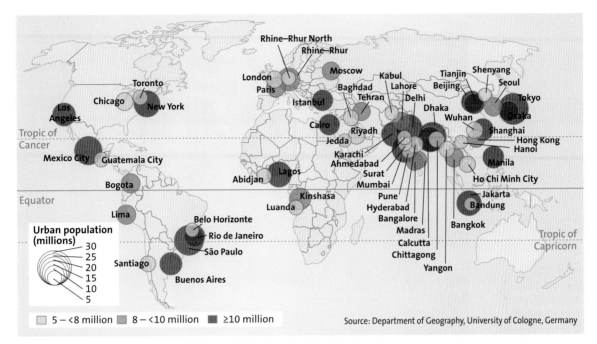

Figure 8.3
The global distribution of megacities

In 2010 there were over 20 megacities with populations greater than 10 million, and nearly 60 exceeding the 5 million mark. Particularly noteworthy is the fact that the majority of today's megacities are located in what we would refer to broadly as the developing world. Furthermore, many have only appeared over the last 25 years. The incidence of megacities in China and India is impressive and mainly reflects the fact that these two countries have huge populations. But megacity mania has also hit much smaller and lower-income countries, such as Colombia (Bogota) and Peru (Lima) in South America, and Angola (Luanda), Congo (Kinshasa) and Ivory Coast (Abidjan) in Africa.

Benefits

This 'mushrooming' of megacities, particularly in developing countries, represents one of the greatest shifts in the distribution of population. While natural increase plays a part in 'promoting' cities into megacities, the altogether more significant population boost comes from rural-to-urban migration on a massive scale and, to a lesser extent, from urban-to-urban migration. These migrations and the associated emergence of huge cities are, for many developing countries, early steps on the development pathway. The new megacities with their expanding range of secondary and tertiary activities and jobs are the catalysts of national economic development. They act as powerful magnets to people and for some, but not all, they offer the opportunity of a high standard of living and better quality of life.

Megacities are also performing a largely unsung role in reducing the rate of global population growth (see *Case study 5*). Although megacities do indeed teem with people, the fact of the matter is that urban birth rates are significantly lower than rural ones. In many megacities, fertility falls to, or even below, the replacement rate of 2.1 children per family. Without the massive rural-to-urban migration, the global population would be growing at a faster rate than it is today. In short, the more the global population

Figure 8.4 Luanda, Angola: (a) poor living conditions in slums; (b) traffic congestion

Jenny Matthews/Alamy

Zute lightfoot/Alamy

shifts to megacities and other large urban areas, the sooner will the point of zero population growth be reached.

Costs

It should be stressed, however, that while megacities offer many opportunities and benefits, they also generate considerable costs. Those costs are both internal and external. Internal costs include:

- Homelessness and slums — the escalating demand for housing far outstrips supply. Many are forced to live in appalling conditions in slums and shantytowns (Figure 8.4 a). Others are unable to access even the most basic form of shelter.
- Traffic congestion — this results from the large-scale agglomeration of people and economic activities, and under-investment in transport infrastructure (Figure 8.4 b). This is but one symptom of the fact that in most developing countries megacities grow faster than their infrastructure.
- Environmental pollution — mainly the pollution of air by factories and motor traffic and the pollution of water by inadequate sewage treatment systems. Other forms of pollution include noise, smell and visual eyesores.
- Vulnerability to hazards — increased by very high population densities.

- Social disorientation — many rural immigrants find it difficult to adjust to a very different lifestyle. This results in stress and alienation, some of which in conjunction with acute poverty may manifest itself in high rates of crime.

The external costs of megacities include:

- Urban sprawl — uncontrolled expansion of the built-up area not only erodes potentially valuable farmland, but also results in high levels of dysfunction and widespread ecological damage.
- Creation of huge peripheries — largely the consequence of the massive rural-to-urban migration. The megacity is a major core draining rural areas of the labour needed for food production, much of which is used to feed megacity people.
- Rural areas and lesser urban settlements are condemned to become part of the vast periphery through being starved of investment and development programmes.
- Social upheaval — families become fragmented as more ambitious members take their chances and move to a megacity.

Despite all these obvious costs, it is understandable that governments in developing countries are reluctant to do anything that might impede the much-needed economic development that is being spearheaded by the megacity. But in the fullness of time the realisation will come that megacity mania and its acute agglomeration of people are unsustainable. Because they are home to acute poverty, social inequality and environmental degradation, megacities have the potential to become centres of global risk.

The emergence of megacities represents a profound shift in a country's distribution of population being activated by domestic migration on a massive scale. Megacity mania is putting many governments in a dilemma: namely, that by taking actions to rectify the inherent costs associated with megacities, they run the risk of thwarting the process of economic development.

It is universally acknowledged that we live in an age of globalisation. The growth of the global economy and advances in transport and communication have greatly increased people's mobility. Never before have so many millions of people wished to, and been able to, migrate, largely in search of a better quality of life. The megacity is testimony to this wish, perhaps mainly in the context of domestic migration. However, the migration wish increasingly has an international dimension.

The wishes of many potential international migrants are being thwarted because some national governments, namely those of popular destination countries, are finding it necessary to control the numbers of immigrants entering their territories. There is an increasing tension between this tightening of immigration controls and what many recognise as a basic human right — freedom of entry. As we made clear in Part 4, there are two sides to every migration — an origin and a destination. In the next case study, the focus is very much on the latter.

THE RIGHT TO ROAM THE WORLD   Case study 35

Freedom of movement is one of a number of basic human rights stipulated in the UN Declaration of Human Rights (1948). The UK is keen to be seen as a law-abiding country, but finds itself increasingly driven to ignore a person's right to roam.

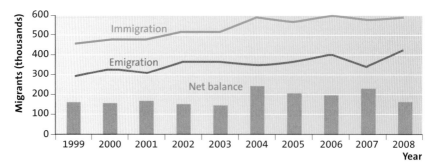

Figure 8.5
International migration into and out of the UK, 1999–2008

During the opening years of the twenty-first century, the annual international migration balances of the UK were positive, with slightly raised net values between 2004 and 2007 (Figure 8.5). However, when the distinction is drawn between British and non-British citizens, two very different situations emerge. The record for the former has been one of persistent net loss, while that of the latter has been one of considerable net gain. British citizens are leaving the country and non-British citizens are arriving. Among the non-British citizens are increasing numbers of asylum seekers.

It is important to remember that this case study is about people attempting to enter the UK from outside the EU. Citizens from member countries (except the Accession 8) have the right to roam within the EU and to settle and work where they wish. See *Case study 19* for more information about the issues raised by these migrants. In particular, refer to Figure 5.11 on page 67 to provide some context.

Between 1982 and 2001, the number of people seeking asylum in the UK showed a 20-fold increase, from around 4,000 to 80,000 per year. The more recent asylum seekers are part of a broad movement of population towards Europe (particularly to the EU) from Africa, the Middle East and Asia. To a lesser extent, they represent migration within Europe away from the trouble-torn former Yugoslavia and the poverty-ridden countries of the former Soviet bloc. The most worrying aspect for the UK is that it is the most popular asylum-seeker destination. The reasons for this are to do with personal perceptions and possibly a reputation that the UK is a 'soft touch'. However, the arrival of such huge numbers of asylum seekers at our borders has raised a fundamental issue — should all asylum seekers be allowed to enter and settle? In order to understand this issue, it is vital to be clear about:

- the difference between an asylum seeker and a refugee
- the immigrant pathways into a country

Asylum seekers and refugees

An **asylum seeker** is a person who seeks to gain entry to another country by claiming to be a victim of persecution, hardship or some other compelling circumstance. A **refugee**, as defined by the UN, is someone whose reasons for moving are genuinely to do with fear of persecution or death. Figure 4.8 on page 52 shows three broad pathways into a country. Particularly important here is the 'conversion' of an asylum seeker into a refugee. Although official statistics are hard to come by, it appears that less than half of all asylum seekers are granted leave to stay in the UK.

Viewpoints

Figure 8.6 looks at the issue of asylum seeking from three different viewpoints currently held in the UK. Those viewpoints — admit none, admit some and admit all — occupy different points along what might be called the 'immigrant welcome scale'.

Who to admit?	Why?	Where to locate?
NONE	Reason 1: the 2001 census shows that while the UK is becoming more multicultural, second- and even third-generation immigrants are still among the most deprived in society (*Case study 10*). The figures show that black, Asian and other ethnic minorities are twice as likely to be unemployed, half as likely to own their own home and run double the risk of poor health, compared with white Britons. The census data undermine the belief that first-generation immigrants may suffer, while their subsequent British-born offspring move off the breadline and become better off. Reason 2: in 2003, an all-party group of MPs claimed that the huge numbers of asylum seekers entering Britain were threatening the country and likely to spark social unrest. They said that the situation had already provoked a political backlash, with voters turning to extremist parties in protest against the spiralling numbers who have come to Britain during the last 20 years. If the current rate of entry persists, the country's capacity to cope, in terms of providing housing, work and social services, would be overwhelmed. Better to close the doors now than run the risk of social unrest.	Send back
SOME Refugees	It is clear that not all asylum seekers are genuine refugees (people whose reasons for moving are to do with a real fear of persecution or death). The fact that so many asylum seekers entering the European Union make for the UK raises the suspicion that many of those claiming to be refugees are in fact 'opportunists' looking for work and a better life. Equally, the fact that most come from Iraq, Zimbabwe, Somalia, Afghanistan and China suggests that persecution may be the key push factor. The challenges are to discriminate between refugee and opportunist; to process asylum applications as quickly as possible; to treat all applicants humanely while this is being done; to ensure that those denied leave to stay are 'removed' and do not 'disappear', swelling the number of illegal immigrants.	Concentrate or disperse
Close family	There are powerful humanitarian arguments for admitting close relatives of people already living legally in the UK. The problem is how far to extend the family net. Should it be limited to next of kin — parents and children — or extended to aunts, uncles and cousins? Then there is the delicate issue of marriage. Should a line be drawn between 'arranged' and 'chance' marriages, and between foreigners entering the country as a national's partner, rather than as a legal spouse?	Where appropriate
Key workers	In 2002 the British government set up the Highly Skilled Immigrant Programme to meet shortfalls in the skills market. The scheme made it easier for the brightest foreign students to carry on working in the UK after their courses finish. Work permits are given readily to international business executives. The former Labour government wanted to admit 150,000 key workers a year as part of the programme. This selective-door process was a reversal of the closed-door policy that had prevailed since 1970. In the 1950s and 1960s, when cheap, unskilled labour was in short supply, the government had operated an open-door policy. The HSIP programme was ended by the coalition government in 2010, which wanted to 'cap' the number of migrants entering the country.	Where needed
ALL	Many people believe that freedom of movement is a basic human right — it was enshrined in the UN Declaration of Human Rights (1948). The view of civil liberties organisations around the world is that this right is paramount. It should be upheld, even to the point of disregarding any political, economic, social or practical problems. However, there are few governments today actively supporting the right, other than when another government is keen to rid its country of a particular group of people or when governments are concerned about the safety of minority groups in another country (e.g. Muslims in Serbia).	Where they choose

Cold → Welcome scale → Warm

Figure 8.6 Asylum seekers in the UK: attitudes and issues

The 'all-are-welcome' viewpoint is based on the belief that freedom of movement is a basic human right, as stated in the UN Declaration of Human Rights (1948). Most subscribers to this view would argue that, migrants having entered a country, this 'freedom' should extend to allow them to choose where they settle. A minority would accept that for practical reasons (availability of housing, work and services) there has to be some degree of intervention in, or management of, location.

The keynotes of the 'some-are-welcome' viewpoint are screening and selection, starting from three different bases:

■ discriminate between the genuine refugee and the opportunist (economic refugee)
■ admit only close relatives of existing citizens
■ admit those who may be expected to bring some 'benefit' or meet a particular demand — specific skills, investment capital and cheap labour

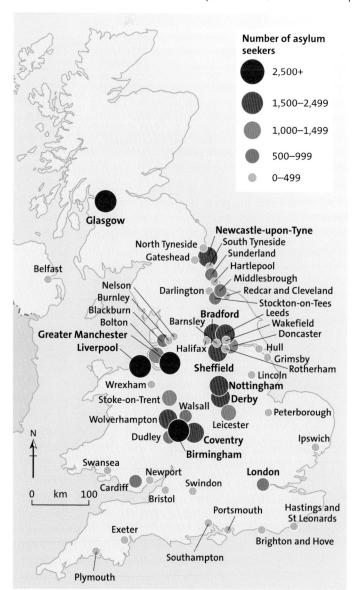

Figure 8.7
Where asylum seekers are sent

Taking the first bullet point, there are two conflicting views about what to do with the genuine refugees — concentrate or disperse them. Considering the second and third bullet points, there would be general agreement that the question of 'where to locate' should be left to family choice and market forces. The result could be either dispersion or concentration.

Arguments for the 'none-are-welcome' viewpoint focus on:

■ the likely plight of immigrants — discrimination, persecution, poverty and hardship, inadequate housing and unemployment
■ the impact of immigrants — on housing, unemployment and overstretched services

Each of the viewpoints on the admission of asylum seekers into the UK (Figure 8.6) can be supported, to varying degrees, by the quotation of official statistics. Each can be made to look persuasive. Who is to say which of these views is right or wrong? In 2010, the new Coalition Government set up a temporary 'cap' of 24,100 immigrants to apply up to April 2011. Beyond that, it intends to set a permanent annual 'cap' of an as-yet-unspecified figure of less than 100,000.

Where to locate?

From a geographical viewpoint, perhaps the most interesting aspect of the issue is 'where to locate?' This relates particularly

Contemporary Case Studies

to two critical stages along the asylum seeker's pathway:

- while the asylum application is being processed
- when permission is granted to stay

Given that most asylum seekers arrive either by ferry at the Channel ports or at London's airports, it is clear that such convergence puts a great strain on reception services, particularly in London and Kent. For this reason, it is now policy to disperse asylum seekers and to hold many of them in designated centres (Figure 8.7). The policy of dispersal seems to have worked reasonably well. However, the scheme has been suspended in certain areas (e.g. Manchester, Burnley and Bolton) at the request of the police and local councils. This has been justified on the basis of nipping threatened ethnic disturbances in the bud. It is proposals to set up large, sometimes purpose-built, holding or detention centres that have sparked the fiercest local resistance.

The issue of location comes to the forefront again once permission has been granted for the asylum seeker to stay. At this stage, the policy of dispersal has, once more, much to commend it. Surely, this is preferable to a policy of concentration, particularly if the intention is that such incomers should be assimilated into Britain's increasingly multiethnic society? Concentration means segregation; segregation means little assimilation.

This case study has only looked at one side of a two-sided situation. Just think of those millions of would-be migrants wishing to escape poverty and persecution who are frustrated by the closing of national doors. Whose best interests should prevail — those of the economic migrant and asylum seeker or those of the destination country?

This issue has many different facets that have not been explored here — facets that are to do with human rights and ethics, party politics and economics. It is important that we should be as fully informed as possible about all aspects of the issue, and not just the geographical dimension that has been brought into focus in this case study.

AN UNEQUAL WORLD

Case study 36

A theme that has surfaced in a number of case studies in this book has been the notion of a world characterised by unequal opportunities. One of the most obvious pieces of evidence of this is the global North–South divide separating the developed and developing parts of the world. In this context, the unequal opportunities are largely to do with differential access to resources, education, capital, enterprise and technology. The outcomes of similar unequal opportunities are found on a smaller spatial scale within individual countries, for example between their core and peripheral regions (see *Case studies 11* and *34*). With both of these examples, the unequal opportunities relate to what is needed to sustain economic development. In both cases, the inequality produces what are commonly known as 'development gaps'.

Wealth

It is not only places and spaces that are involved in today's unequal world. People are also very much involved. The most obvious and widespread inequality is between rich and poor people. At least 80% of the world's population live on less than $10 a day (Figure 8.8). The poorest 40% of the world's population account for a mere 5% of global income. The richest 20% of the world's population account for 75% of world income.

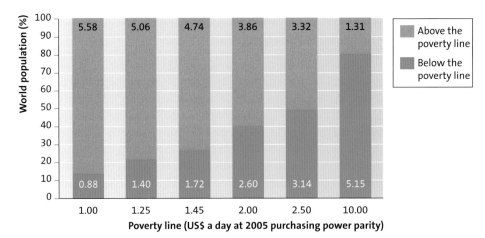

Figure 8.8
Percentage of global population at different levels of poverty

Note: Numbers inside bars are world population at that indicator (billions)
Source: World Bank Development Indicators 2008

More than 80% of the world's population live in countries where income differentials are widening. This widening of the gap between rich and poor is truly an indictment of the modern world.

Ethnicity

Ethnicity provides another powerful basis of inequality. In many of the world's multi-ethnic societies, it is frequently the case that one ethnic group happens to command more power and influence than the others. Such a situation can quickly lead to exploitation, discrimination and, in extreme cases, persecution. The population of Iraq, for example, is made up of two main groups: the Shia (65% of the population) and the Sunni (32% of the population). Despite the numerical difference between the two groups, it is the Sunni who have been the more powerful. Saddam Hussein was a Sunni and during his reign of terror was responsible for the persecution of the Shia. Since his removal following the US invasion of Iraq, hostility between the two groups has not abated. Indeed, both sides continue to indulge in a long succession of revenge killings. The Sunni use suicide bombers in public places to strike at the more numerous Shia. The Shia have death squads that target prominent Sunnis. US attempts to create a new power-sharing government have yet to bear fruit. The chances of ever achieving an integrated society with equal opportunities are minimal.

Recent history is scarred by incidents of **ethnic cleansing**: the Nazi attempt to exterminate the Jews in Europe; the atrocities that followed the break-up of the former Yugoslavia; the genocide in Rwanda. There seems to be no easing in the frequency of such horrific incidents. They all fly in the face of the vital humanitarian principle that all people are equal.

Gender

A final manifestation of today's unequal world relates to gender. While most developed countries are now wholly committed to the principle of gender equality, there remain extensive areas of the world where women continue to be treated as second-class citizens. Women often have less access than men to legal recognition and protection, and to public knowledge and information. They have less decision-making power both

Figure 8.9
Women attending a literacy class in Meru, Kenya. Despite improvements, illiteracy among women is still a major cause of poverty

TopFoto

within and outside the home. They have little control over fertility, sexuality and marital choice. What they wear and do is decreed. They are denied freedom of choice.

Oxfam, which campaigns vigorously throughout the world for women's rights, has highlighted the following key facts of gender equality:

- In 2009 women on average accounted for less than 18.4% of members of parliament. At all levels and in all sectors, fewer women than men are part of decision-making processes.
- Over two-thirds of the world's 776 million illiterate people are women and, despite improvements, more than 55% of the 75 million primary-age children who do not go to school are girls. Illiteracy among women is a major cause of poverty (Figure 8.9).
- Worldwide, women earn on average only 84% of what men earn in formal waged work. However, large numbers of women are concentrated in informal and precarious work, associated with low and unstable earnings.
- Every year over 500,000 women die of pregnancy-related complications, and between 8 million and 20 million a year suffer serious injury or disability from the same causes.
- Between 10 and 70% of women report abuse by their intimate partner in every country where reliable data exist. Systematic rape has left millions of women and adolescent girls traumatised, pregnant or infected with HIV.
- 80% of the world's 35 million refugees and **internally displaced persons** (IDPs) are women and children. Men often represent the majority of casualties in conflict. However, in most humanitarian crises, women's vulnerability increases, as do the difficulties associated with reproduction.

This case study has drawn attention to a few aspects of today's unequal world. They may be only the tip of the iceberg. Particularly sobering is the likelihood that, despite centuries of so-called progress and development, the world is becoming more unequal.

Examination advice

The case studies that have been an integral part of your A-level geography course can be put to good use in a variety of ways in the examination. Much depends on the following:

- The question task and the command used in the wording of a question. Table 9.1 indicates that there are at least four different question scenarios: name, support, compare and examine. They are ranked in terms of what is expected in the degree of case study detail, which ranges from simple name-dropping to detailed knowledge and understanding of a particular situation.
- The context of the question. The challenge here is to find an example or case study that is appropriate to the question topic. The chapter contents (see pages 3–5) should help you here, because the 36 case studies presented in this book are organised according to the main topic areas of population geography.

Table 9.1
Question commands and the examiner's case study expectations

Required case study detail

Task or command*	Name	Support	Compare	Examine (a particular situation or statement)
Typical question	Give an example of a country at Stage 1 in the DTM.	With the use of examples, explain why the relationship between population and resources is so important.	Evaluate the birth control programmes of two named countries.	With reference to a named country or region, examine the main factors affecting the distribution of population.
Case study expectation	No more than the name of an appropriate country, e.g. Ethiopia, Bangladesh	Use of at least two contrasting case studies, e.g. an overpopulated and an underpopulated country; more than name-dropping needed	Better to opt for two contrasting case studies, e.g. India (voluntary; limited success) and China (enforced; successful)	A sketch map showing the salient features of the distribution is a minimum requirement; a systematic look at key factors should follow

*A word of caution — watch out for those questions that do not specifically ask for examples, but nonetheless expect them. For example:

- Suggest reasons why some migrations involve hazardous journeys.
- 'Migration creates as many problems as it solves.' Discuss.
- To what extent do you agree with the view that the global population problem is not one of numbers, but of distribution?

If you are in doubt as to whether examples are required in your answer to a question, it is better to give some rather than none.

In the rest of this section, we look at five different examination contexts that require the use of case studies. There are the four shown in Table 9.1; they apply mainly to unseen examination papers. The fifth is where the geography specification requires you to undertake an inquiry into a set topic and to submit a report by a given deadline. Therefore, the contexts range from simply naming an example of a particular situation, through using case study material for support purposes, to an extended and detailed use of case studies.

Naming examples

This requires nothing more than being able to cite one of your case studies as a relevant example. The test here is one of appropriateness. In part (b) of the sample question below, it would not suffice to name just any densely populated country, such as the Netherlands or Singapore.

19 Question

(a) Identify three symptoms of overpopulation.
(b) Name one country that is widely recognised as being overpopulated. Give reasons for your choice.

Guidance

(a) Very high population densities; malnutrition or starvation; widespread poverty; high incidence of disease; environmental stress
(b) China/Indonesia/Puerto Rico/Haiti
The reasons are to do with the symptoms shown by the chosen country. Does it show all the symptoms, or only some?

Naming and using a supporting example

The next level up requires that you do more than just name an appropriate example. You are also expected to demonstrate knowledge and understanding by providing some detail.

20 Question

With reference to a named developed country:
(a) identify a region of low population density
(b) outline the physical factors that help to explain that low density
(c) suggest one economic development that might help to raise population density

Guidance

(a) The island of Hokkaido (Japan)
(b) ■ Harsh climate, particularly in winter — inhospitable living conditions
   ■ Prevalence of mountainous terrain and thin soils
   ■ Remoteness from the 'core' of Japan
(c) The promotion of tourism — winter sports and wilderness adventure activities

Using case studies comparatively

Essay-type questions that require the comparative use of case studies are popular with chief examiners. Much of the challenge of such questions hinges on selecting appropriate case studies. In some instances, the choice is fairly obvious and restricted. In others, there may be more choice and options than you might first think from an initial reading of the question.

The trouble with all examination questions that ask you to compare situations (as represented by appropriate case studies) is that some candidates believe that all they have to do is to rehash each study in turn. This leaves examiners to draw their own conclusions as to whether or not the situations are similar. In short, the question is not answered and relatively few marks can be gained. In planning effective answers to the comparative type of question, it is necessary to interleave references to your chosen case studies. Look back at *Using case studies 15* (page 88) for an illustration of how this might be done.

21

Using case studies

Question

With reference to two contrasting examples, evaluate the role of government policies in influencing rates of population change.

Guidance

The obvious choice here is to compare a country that has tried to curb its birth rate with one that has done the opposite. Romania (research this for your own case study) and China (*Case study 29*) are two immediate candidates. Equally, you would be justified in choosing China and India as examples of government intervention with two different outcomes — success and relative failure respectively.

Take a closer look at the question. There is nothing here to say that your answer can only be about policies to do with birth rates. Any country that has positive policies about health and medical services stands to affect rates of population change through increased life expectancy and reduced mortality rates. So, you could build an answer around a comparison of the UK (*Case study 7*) and Russia (*Case study 8*).

When you take this closer look at the question, you will perhaps begin to realise that there is nothing in the wording to say that you should not focus your answer on migration rather than natural change. A discussion involving a comparison of an 'open-door' country such as Australia (*Case study 30*) or Spain (*Case study 31*) today with a 'restricted entry' country such as the UK (*Case study 35*) would allow you to pick out some contrasting demographic changes with respect to birth and death rates, age structure and ethnicity.

Building an essay around a single case study

A successful attempt to answer any question that starts 'With reference to a named country...' requires three things:
- choice of an appropriate country
- sufficiently detailed knowledge of that country in relation to the particular question
- resisting the temptation to set down all that you know about your chosen country (the 'everything but the kitchen-sink' approach), but instead harnessing only those aspects that are directly relevant

In addition, citing a few appropriate statistical facts can help give your essay focus and conviction.

Question

For a named country, examine the demographic, economic and social consequences of a declining fertility rate.

Guidance

Given the case studies presented in this book, there is much to commend the UK as your choice (*Case studies 1* and *7*). You might find some prompts in *Case study 33* about the sorts of consequences that should be covered.

Demographic consequences include:

- a fall in the rate of population growth
- 'greying' of the population structure
- a shift in the nature of dependency from the young to the elderly
- encouragement of immigration to make good the shortfall in births

Economic consequences include:

- falling unemployment and labour shortages
- a rising per capita tax burden to pay for services increasingly oriented towards the elderly
- changing consumer demand
- a shift in service needs — more residential homes for the elderly; fewer schools

Social consequences include:

- a decline of youth culture
- changing values in a 'greying' society
- innovation giving way to experience
- fewer mothers and more career women

Planning an extended essay involving a range of case studies

The standard advice on essay planning should be followed: namely, that the essay should have a three-part structure — a brief **introduction** and an equally brief **conclusion** (one paragraph each), separated by a series of paragraphs (the **expansion**) that develop your argument or discussion points. It is here that you should incorporate supporting examples and case studies.

Grasping and applying the advice contained in this part of the book should have a positive outcome. It is sometimes easy to forget that geography is about the real world. The more contemporary examples that you can include in your examination work, the more you are likely to convince and impress the examiner that you have a sound knowledge and understanding of today's world. A good dose of reality in the form of relevant case studies and examples can work wonders when it comes to raising AS and A2 geography grades.

Question

Which do you think poses the greater challenge to a country — a high rate of natural increase or a large influx of immigrants? Justify your viewpoint with reference to examples.

Guidance

Introduction

Rather than jumping in at the deep end and naming your choice, you might be a little more subtle. You could start by making the point that since both scenarios are likely to result in rapid population growth, they therefore pose the same broad challenges.

Expansion

First, look at those common challenges. They include more mouths to feed, increased pressure on resources, need for more housing, raised demand for services, and the spatial distribution of the resulting population growth.

Now move on to pose the question in a slightly different form — given these similarities in outcome, are there any differences? If so, are these differences significant when it comes to government policies aimed at controlling each of the two situations?

Now look at the outcomes and challenges that are specific to each of the two scenarios. With respect to natural increase, there will be a need for more child-oriented services, such as clinics, midwifery and schools. Contrast a developing country (e.g. the Gambia or French Guiana), where the resources simply do not exist to meet this particular demand, with say one of the oil-rich states (e.g. Kuwait or UAE). With respect to immigration, outcomes and challenges include:

- immigrant characteristics (e.g. ethnicity, skills)
- attitudes and perceptions of the native population
- integration into society through education and equal opportunities

Once again, the UK would serve well as an example (*Case studies 20* and *35*).

Conclusion

Large-scale immigration poses more immediate challenges. However, a high rate of natural increase could be more threatening: first, because most countries experiencing it rank among the least developed; second, because the time-lag between taking action (i.e. introducing a birth control programme) and seeing any significant impact on the rate of natural increase is likely to be a long one. In contrast, it is possible to close the 'immigration door' fairly swiftly.

Appendix

Finding and organising case studies

This book contains 36 case studies that can be used to illustrate and support the key ideas relating to population and migration. However, you may choose to find some of your own. To do this, you might tap into the following sources:

■ A-level textbooks — where the case studies are specially chosen
■ magazines and periodicals — such as *Geography Review, Geofile, Geo Factsheet* and *Geo-News Review*
■ videos, CD-ROMS and television programmes — you will need to take notes; television programmes have the advantage of being up to date; iPlayer may be useful here
■ the internet — the opportunities are overwhelming; you will need to restrain your searches. Websites that currently provide useful information and material about population and migration are listed below

> **www.census.gov/ipc** is the US census bureau site — you can find population pyramids for any country in the world by going to **www.census.gov/ipc/ www/idb/informationGateway.php**

The following three websites are good sources of global and national population statistics:

> **http://data.worldbank.org/topic**

> **http://unstats.un.org**

> **www.cia.gov/library/publications/the-world-factbook/index.html**

> **www.statistics.gov.uk** is the source of all official UK data

> **www.visionofbritain.org.uk** includes maps, statistical trends and historical descriptions

> **www.unfpa.org** gives you access to the state of the world population — and **www.unfpa.org/pds/trends.htm** contains all sorts of population information

> **www.overpopulation.com** and **www.populationinstitute.org** look at the issues of population and resources, but beware of bias on this issue

> **www.nidi.knaw.nl/web/html/pushpull** provides a range of basic migration information

> **www.worldrefugee.com** gives you an insight into current refugee movements

■ your local area — local newspaper, local authority plans and statistics

Case study data are best kept in a succinct note form, perhaps on filing cards, with key facts and figures set out as bullet points. Alternatively, much can be summarised by annotating a sketch map or diagram. Always note the date and source and also look for bias.

As you collect your case studies, you should crosscheck them against the content of your AS and A2 geography specification. You need to ensure that you have good and even coverage of:

- the key ideas
- countries and places at different stages of development and at different spatial scales

You should also be aware that some case studies could be used to support more than one of the key ideas.

Index